Iraqi Fighters, 1953-2003: Camouflage & Markings

By Brig. Gen. Ahmad Sadik & Tom Cooper

IRAQI FIGHTERS
1953–2003: Camouflage & Markings

By **Brig. Gen. Ahmad Sadik** & **Tom Cooper**

HARPIA
PUBLISHING+

Copyright © 2008 Harpia Publishing, L.L.C. & Moran Publishing, L.L.C. Joint Venture
2803 Sackett Street, Houston, TX 77098-1125, U.S.A.
iraqifighters@harpia-publishing.com

All rights reserved.

No part of this publication may be copied, reproduced, stored electronically or transmitted in any manner or in any form whatsoever without the written permission of the publisher.

Lay-out & typesetting by Studio D&M, Zagreb, Croatia

Printed at Tiskara Zelina, Croatia

Harpia Publishing, L.L.C. is member of

ISBN 978-0-615-21414-6

Contents

Introduction	7
Acknowledgements	8
Abbreviations	9
Origins of RIrAF & IrAF National Markings	11
Origins of RIrAF & IrAF Unit Insignias	12
Special Insignia	12
Original Names of Iraqi Air Bases	13
Chapter 1: deHavilland Vampire & Venoms	15
Chapter 2: Hawker Hunters	23
Chapter 3: Mikoyan i Gurevich MiG-17s	37
Chapter 4: Mikoyan i Gurevich MiG-19s	45
Chapter 5: Mikoyan i Gurevich MiG-21s	49
Chapter 6: Mikoyan i Gurevich MiG-23s	71
Chapter 7: Mikoyan i Gurevich MiG-25s	81
Chapter 8: Mikoyan i Gurevich MiG-29s	89
Chapter 9: Dassault Mirage F.1s	95
Chapter 10: Sukhoi Su-7s	117
Chapter 11: Sukhoi Su-20/22s	123
Chapter 12: Sukhoi Su-24s	137
Chapter 13: Sukhoi Su-25s	141
Appendix: I. Bibliography	147
II. Kit & Decal Lists	148
Index	151

Introduction

"Iraqi Fighters, 1953-2003; Camouflage and Markings", came into being in the course of attempts to organise the results of research about the history of the former Royal Iraqi Air Force and Iraqi Air Force. Yet, development and production of this book were necessitated by problems encountered by its authors, which we feel deserve to be described in some detail.

Originally, we had hoped for the situation in Iraq to normalise to one degree or another, and make a thorough academic research of the IrAF history possible. Massive amounts of documents and photographs used to be available in Iraq, both in official archives, and private collections and we expected that one day it would be possible to work ourselves through these, with corresponding results.

Sadly, the situation in Iraq is such that most of official archives have been completely destroyed. Documents, photographs and other evidence from private collections are regularly lost, and most of original aircraft that survived the war of 2003 have been reduced to heaps of wreckage. Many of former Iraqi Air Force officers, pilots and serviceman are scattered around the world, many are in hiding while others pursuing different careers. The longer it takes to research, the less documentation is available.

Materials we have gathered so far, varied a lot in quality. Even if only very few poor photos from the period of 1950s and 1960s became available, the records are almost complete. We were able to find out most about the number of aircraft acquired during these times, and reconstruct almost the entire RIrAF and IrAF serial numbering system, unit insignia and other similar details.

Subsequent times, especially those during the war between Iran and Iraq, were much more problematic to cover. The number of available photographs is lower, and very few official documents available. In a few places we have therefore had to reach back on fading memories of former IrAF officers, as well as foreign documents and publications. We have done our best to sort these out, and provide the most reliable information.

Another major problem of publishing a book of this kind is that of finding a publisher. Various projects related to Arab air forces have been offered by authors and a number of their co-workers to several prominent publishers about military aviation history in the last years. All have been turned down; the usual explanation being that the editors (or owners) do not believe there is sufficient interest in this topic. Yet other publishers would not accept our work because we do not have a large number of high-quality photographs. It is a sad fact that in our times there are many responsible persons, but also plenty of potential readers, who tend to forget about the situation in Iraq, or the fact that not all the air forces have the same approach to public relations like those in the West.

Eventually, we have been forced to organize the publishing practically on our own, while left in a situation where there is plenty to say about the IrAF. There is little to show in terms of quality photographs. The exact amount of information waiting to be presented is the next problem we have had to tackle, one connected to modern-day publishing. Writing big, hefty volumes, packed with valuable information takes plenty of time. It also costs a lot to produce, and tends to result in very expensive products.

Studying the potential market, and reaching back on experiences from various other publishers, we have therefore decided to try to prepare a series of smaller volumes instead, this being the first one. This way is promising to offer us the chance of concentrating on very specific topics, and providing much more in-depth information. For such reasons, this relatively small volume is dealing almost exclusively with the Iraqi fighter jet aircraft, their camouflage and markings, and less with "people" of this once proud service. We apologise in advance to any reader that might feel offended by this approach and promise to do our best to cover the other aspects to much greater detail in the future.

Why "1953-2003"? Preferably, we would have started with a volume on the history of Iraqi Air Force from its inception, in 1931, its destruction in 1941, and then go on with subsequent periods in a similar manner. This extremely problematic situation with available materials, as well as potential publishers, has forced us into "starting in the middle". As of yet we are unable to prepare a volume on other related topics or other periods since we lack photographic material, and hopes of attracting sufficient attention from the public are low. Therefore, we decided to follow the most promising idea: that of preparing this work on jet fighters in service with IrAF between 1953 and 2003.

This book is thus based on a series of compromises. Nevertheless, we hope that the readers are going to find it a valuable source of reference for reliable information on IrAF. At least we are sure that this is the first time than an Arab air force is presented to comparable depth and detail.

Our hope is also that sometimes in the future, a much better volume about the history of IrAF will be possible.

Brig. Gen. Ahmad Sadik (IrAF) & Tom Cooper, January 2008

Acknowledgements

It is a great pleasure of both authors to record our thanks to the people who contributed to this book. First and foremost, we would like to thank to all the former IrAF officers and pilots that kindly shared their recollections and documentation with us. Sadly, due to the current situation in Iraq, we do not feel free to reveal their names in public.

We have to express our gratitude to a donor of over 500 photographs of various IrAF aircraft taken in Iraq since 2003, who prefers to remain anonymous. This book would be unimaginable without your help.

Several people have supported, read, commented on, and encouraged our efforts over the time. Strong influence and traces of these wonderful friends, foremost David Nicolle, Farzad Bishop, and Ugo Crisponi, are imprinted on each page of this book. Without kind help from Ferencat Vajda it would have been impossible to put together a list of IrAF serials for the 1950s and 1960s. Several gentlemen from IPMS Austria, foremost Christof Hahn, provided valuable help researching about model aircraft, but also in regards of book design. Finally, Tom Long and Yaser al-Abed, who have always supported our work with all their powers, should be mentioned as well as Adrian DeVoe, who helped so much with editing the manuscript. Our special thanks to every one of them.

Abbreviations

AAA	Anti Aircraft Artillery
AAM	Air-to-Air Missile
AB	Air Base
AIM-120	Air-Intercept Missile-120 AMRAAM
APC	Armoured Personnel Carrier
ARM	Anti-radar-missile
ASCC	Air Standardization Coordinating Committee
AWACS	Airborne Warning and Control System
CAP	Combat Air Patrol
CAS	Close Air Support
CBU	Cluster Bomb Unit
c/n	construction number
DoD	US Department of Defence
Doppler	radar making use of shift in frequency of signals reflected from ahead of or behind aircraft (to give measure of true groundspeed) or of signals received from fixed (Earth) and moving targets
ECCM	Electronic counter-countermeasures
ECM	Electronic countermeasures
EAF	Egyptian Air Force
EO	Electro-optical
ESM	Electronic support (for surveillance) measures
EW	Electronic warfare
FOIA	Freedom of Information Act
GP	General-purpose (bomb)
I-band	EM radiation 8 to 10 GHz
IDF/AF	Israeli Defence Force/Air Force
IFF	Identification Friend or Foe
IIAF	Imperial Iranian Air Force
ILS	Instrument landing system
IR	Infra-red, EM radiation longer than deepest red light sensed as heat
IrAAC	Iraqi Army Air Corps
IrAF	Iraqi Air Force
IRIAF	Islamic Republic of Iran Air Force
IrN	Iraqi Navy
Kh-23	Soviet tactical air-to-ground missile (ASCC Code: AS-7 Kerry)
Kh-25	Soviet tactical air-to-ground missile (ASCC Code: AS-10 Karen)
Kh-28	Soviet tactical air-to-ground missile (ASCC Code: AS-9 Kyle)
Kh-29	Soviet tactical air-to-ground missile (ASCC Code: AS-14 Kedge)
Kh-66	Soviet tactical air-to-ground missile (ASCC Code: AS-7 Kerry)
KIA	Killed in Action
KLu	Koninklijke Luchtmacht (Royal Netherlands Air Force (RNLAF)
kW	Kilowatts, units of 1,000 watts of electrical power
LGB	laser guided bomb

MANPAD	Man-Portable Air Defence system. A "shoulder mounted" SAM that can be operated by a single soldier.
MHz	Megahertz, millions of cycles per second
MiG	Mikoyan and Gurevich
MRAAM	medium-range air-to-air missile
MTI	Moving-target indication, radar can eliminate returns from all sources except moving targets
Nav/attack	used for navigation and to aim weapons against surface target
PAF	Pakistan Air Force
Pallet	Platform upon which one or more mission equipments is mounted for installation in or beneath aircraft
Passive	non-emitting
PD	Pulse-doppler radar
PGM	Precise Guided Ammunition
PL-2	Chinese air-to-air missile (ASCC Code: AA-2 Atoll)
Pod	streamlined container for equipment carried outside aircraft
PRF	Pulse-repetition frequency
PRGM	Russian landing system
R-3	Soviet air-to-air missile (ASCC Code: AA-2 Atoll)
R-13	Soviet air-to-air missile (ASCC Code: AA-2 Atoll)
R-23	Soviet air-to-air missile (ASCC Code: AA-7 Apex)
R-24	Soviet air-to-air missile (ASCC Code: AA-7 Apex)
R-27	Soviet air-to-air missile (ASCC Code: AA-10 Alamo)
R-40	Soviet air-to-air missile (ASCC Code: AA-6 Acrid)
R-60	Soviet air-to-air missile (ASCC Code: AA-8 Aphid)
RAF	Royal Air Force
RHAWs	Radar homing and warning system
RIrAF	Royal Iraqi Air Force
RJAF	Royal Jordanian Air Force
RS-2	Soviet air-to-air missile (ASCC Code: AA-1 Akali)
RSBN	Russian Short-Range Air Navigation System, equivalent to Western TACAN
RWR	Radar Warning Receiver
SAM	Surface-to-air missile
SARH	semi-active radar homing
Semi-active	not itself emitting but homing on radar or other signals reflecting from target
SLAR	Sideways-looking airborne-radar
Su	Sukhoi
SyAAF	Syrian Arab Air Force
UAV	unmanned aerial vehicle
UHF	Ultra-high frequency, 300MHz to 3GHz
VHF	Very high frequency, 30 to 300MHz
VOR	VHF omnidirectional range

Origins of RIrAF & IrAF National Markings

The triangle used as national marking on Iraqi military aircraft is a stylised representation of Iraq, with a green field outlined in black, two rivers (Euphrates and Tigris) represented in form of the red letter "Jeem", which stands for the word "Jaish" (Arabic for "Army"), and a white "diamond"

Ever since establishment of the Iraqi Air Force, on 22 April 1931, national colours were usually applied on the rudder, and consisted of vertical strips in green, white, red, and black. Following the reorganisation of the IrAF in the mid-1940s, a fin flash with same colours and in the same order was introduced instead.

Immediately upon the coup of 1958, the old RIrAF fin flash was retained, since the new flag had the same colours, but a yellow disc added over white and red fields. It was only sometimes in the spring of 1959 that an entirely new fin flash was introduced instead, consisting of vertical strips of black, white and green, with an eight-pointed star on white field, with yellow centre. Following a unity agreement between Egypt, Iraq and Syria, on 17 April 1963, a new fin flash was introduced, in so-called "Pan-Arabic" colours, with horizontal stripes in red, white and black: red for courage in battle, white for generosity, and black for the era of Caliphates and past glory of the Islamic prophet Mohammed. Though three eight-pointed stars outlined in green are usually said to have been applied on the white field, no such flags are known to have been officially issued to any authority in Iraq, and it is unlikely that any were ever applied on IrAF aircraft.

For the Iraqis, there was never a doubt about the meaning of three green stars on the white field of their new flag: the aspiration for unification with Egypt and Syria. The later was definitely represented by the national flag and corresponding fin flash introduced on 28 August 1963. This kept the same three colours, but introduced three green five-pointed stars on the white field.

In this form, it looked practically the same as the fin flash of Syria from 1963 until 1972, and Egypt from 1958 until 1972. This flash remained in service until 1991, and was only slightly changed on 13 January 1991, through addition of "Takbir" ("Allahu Akbar" – "God is Great") in green on the white field.

Artwork: IrAF Roundels

Iraqi national insignia as used by RIrAF and IrAF in the period 1953-2003, from top to the bottom:
- Identification Triangle (applied on all RIrAF and IrAF aircraft since 1931)
- Original fin flash, as in use from 1931 until 1958
- Temporary fin flash, introduced immediately following the coup of 1958
- Two versions of fin flash in use between 1958 and 1963
- Fin flash in use from 1963 until 1991
- Fin flash with "Takbir", in use since 1991

"roundel"

1931-1958

1958-1959

1959-1963

1963-1991

1991-today

Origins of RIrAF & IrAF Unit Insignia

Being a pilot of Iraqi Air Force meant being somebody special, extraordinary, and distinguished. IrAF pilots were held in high regards and enjoyed a special status in Iraqi society. Correspondingly, they proudly wore insignia that presented them as such on their military uniforms. A tag with the rank and name of the person was usually worn above the right chest pocket, and pilot's wings on the left side. Patches with unit insignia were worn on right sleeve of pilot's suit. Aside from being worn on uniforms, unit insignia was always applied at the entrance to squadron ready rooms, mess- and briefing rooms. Several cases are known where either pilots or ground personnel were so happy to belong to a specific unit, that they would paint large unit insignia inside aircraft shelters as well.

The origins of Iraqi unit insignia reach back to times when IrAF was strongly influenced by the Royal Air Force. Iraqi military heraldic in general meant to present strength, power, and nationalistic values, as well as geographic locations. After 1968, Pan-Arabic ideas became prominent, as can be seen by addition of maps of Arab world, or Iraqi and Palestinian flags to insignia of several units.

Despite such general tendencies, specific units still followed their own traditions. No.1 Squadron, for example, proudly presented the mountains of northern Iraq on its patch; No.96 Squadron was more than proud to show the plane it operated (MiG-25); several units (especially ground-attack ones) usually had citations from holy Koran on their patches.

Nevertheless, the practice of applying squadron insignia on aircraft was rather unusual, even if not unknown. A small number of units, especially training assets, have had most of their aircraft adorned with squadron patches, but in most of other cases none were applied at all.

Special Insignia

Time and again through its history, some IrAF aircraft were decorated with special insignia. In most cases, this insignia indicated some kind of a special status of an aircraft type, or specific airframe. In others, it was commemorating a specific event or achievement by the ground crew.

One of the best known, even if most often misinterpreted "special "insignia applied on IrAF aircraft was a small shield in national colours (red, white and black), with white title "Iraq", in Latin and Arabic alphabet, applied on the red field. Such shield was usually outlined in yellow and white.

While for years thought to have been a decoration for particularly well-maintained aircraft, or even unit marking, this marking is now known to have been applied on Iraqi Su-20s and Su-22s only, identifying their special status – in the sense of their importance and excellence – for the IrAF and the whole Iraqi nation.

Another insignia that was sighted in the mid-1990s for the first time was showing a head with barrette, applied in white. This was the marking of "Saddam Fedayeen", applied on aircraft in commemoration of some specific achievement – mainly of technical nature (regardless if for excellent maintenance or successful modification). Saddam Fedayeen insignia was observed on at least one IrAF Mirage F.1EQ – in addition to several Mi-8s flown by Iraqi Army Air Corps.

The third kind of special marking was discovered on several Su-25s only, and first sighted only after these aircraft were captured, in 2003. This consisted of two circles in white, a golden dome and a light blue building: a stylised representation of the "Dome of the Rock" (the famous mosque in Jerusalem). The marking commemorated the participation of specific aircraft in the big military parade in Baghdad, on 31 December 2001.

So called "kill markings" have been worn as well, at latest since the 1967 War with Israel, and only for the purpose of commemorating air-to-air kills.

At the time, Hunters that scored kills against Israeli fighters were decorated by small Stars of David in white, with inscription mentioning the rank and name of successful pilot beneath. During the war with Iran, kill markings showing Iranian roundels were introduced, and applied on several MiG-21MFs and MiG-21bis, MiG-23MFs, and Mirage F.1EQs. No examples of such markings being "underlined" by the name of successful pilot are known.

Original Names of IrAF Air Bases

Due to lack of knowledge about the IrAF, foreign authorities tended to use makeshift names for most of Iraqi airfields. Original names were practically unknown, and even if, then often misspelled. In this work, we have decided to use only the original names of Iraqi air bases, as these were used by IrAF personnel and in documents. For better orientation, below is a list of comparative names as known outside Iraq, and official names of IrAF air bases and airfields. There are two exceptions from this rule. H-3 was not an official IrAF air base at the time of the war with Israel, in 1967, and as such had no official name. Similarly, the former RAF Habbaniyah air base became a part of the Tammuz complex in the 1970s, before that time the Iraqis used its old name from the RAF-times.

Table 1: IrAF Air Bases

US/British Name	Official IrAF Name	Remarks
Al-Assad	Qadessiya AB (former al-Baghdadi AB)	Former water-place Ain al-Assad, named after a battle in which the Arabs defeated ancient Persians, circa. 600AD
Al-Taqaddum	Tammuz (or Tahmouz) AB	Named after the Babylonian month of Tammuz, 7th month in the year
Balad	al-Bakr AB	Named after former Iraqi President, Ahmad Hassan al-Bakr
Balad SE	Dulu'ya	Dispersal facility, named after geographic location
H-2	Sa'ad AB	Named after an ancient Islamic warrior, from cca. 600AD
H-3	al-Wallid AB	Named after an ancient Islamic warrior, Khalid bin al-Wallid, from circa. 600AD
Jaliba	Jaliba	Dispersal facility
Kirkuk	al-Hurriyah AB	Means "Liberty" in Arabic
Mosul	Firnas AB	Named after a legendary middle-age Andalusian Arab, who attempted to fly using feathered wings
Mudaysis	Talha AB	Dispersal facility named after ancient Islamic warrior
Rashid	al-Rashid AB	Named after Islamic Caliph under which Baghdad prospered
Shoaibah (RAF Shaibah)	al-Wahda AB	Means "Unity" in Arabic
Kut or Kut al-Hayy	Abu Ubaida al-Jarrah AB	Named after an Islamic warrior from circa. 600AD
Tallil	Ali Ibn Abu Talib AB	Named after an Islamic warrior from circa. 600AD
Ar Rumaila SW	Artawi airfield	Named after geographic location
Qayyarah West	Saddam AB	Named after former Iraqi President

Chapter 1

deHavilland Vampire & Venom

Vampire FB.Mk.52 & T.Mk.55
Venom FB.Mk.1 & FB.Mk.50

Service History

The first jets to enter service with the Iraqi Air Force were deHavilland Vampires. The original Iraqi order was for 12 aircraft. This included six two-seaters and six single-seaters, all equipped with Goblin 3 engines. The first to be delivered were three T.Mk.55 two-seaters, and six FB.Mk.52 single-seat fighter-bombers. All of which arrived during May 1953. On 13 August 1953 these aircraft officially entered service with the re-established No.5 Squadron, then based at Rashid AB, in southern Baghdad.

Six additional FB.Mk.52s and nine T.Mk.55s (one of which crashed during the delivery flight), followed in 1955, bringing the total to 12 FB.Mk.52s, and ten T.Mk.55s.

In spring 1954, Rashid AB was flooded by the Tigris River, and the whole No.5 Squadron was evacuated – first to Baghdad "civilian" airport and on 10 May to Habbaniyah AB, then still under British control. When the British have left, in May 1955, Habbaniyah became the new permanent base of the Iraqi Vampires.

Useful photographs of IrAF Vampires remain scarce and poor in quality. This one shows details of camouflage on the front of the fuselage, as well as parts of under-wing markings – including the serial number 389, applied very much in typical RAF fashion.
(Ahmad Sadik Collection)

15

Iraqi Fighters, 1953-2003

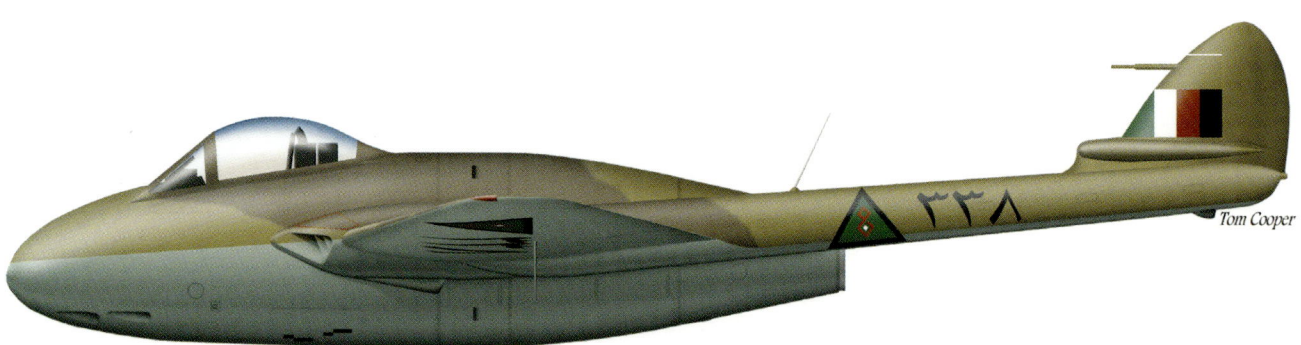

As originally delivered to IrAF, Vampire FB.Mk.52s should have been painted as shown on example of #338, with the border between camouflage pattern of the upper side and Sky Type S colour of the lower side going along the mid-fuselage. Sadly, no pictorial evidence is available to confirm this.

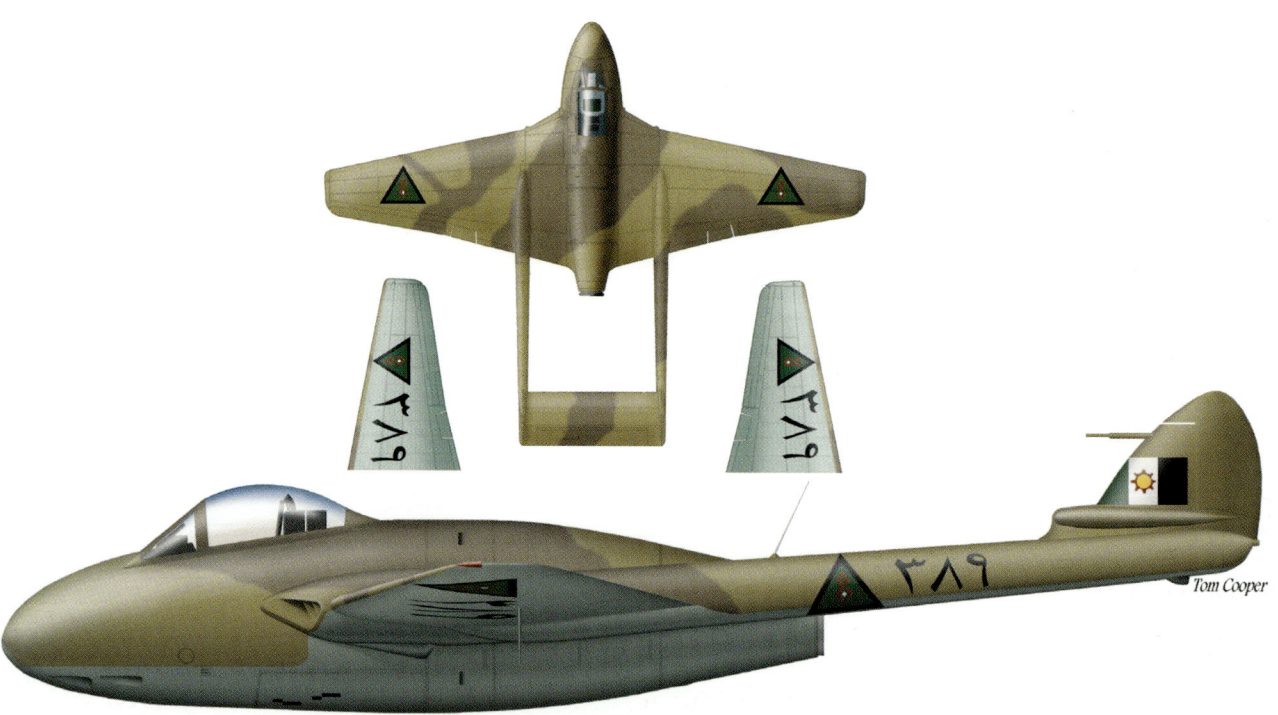

Practically all the available pictorial evidence shows Iraqi Vampires having their front sides painted as shown on FB.Mk.52 #389, with the upper border of Type S colour quite low down the fuselage.

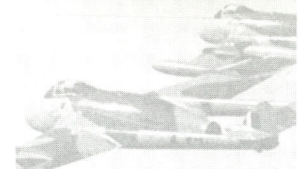

deHavilland Vampire & Venom

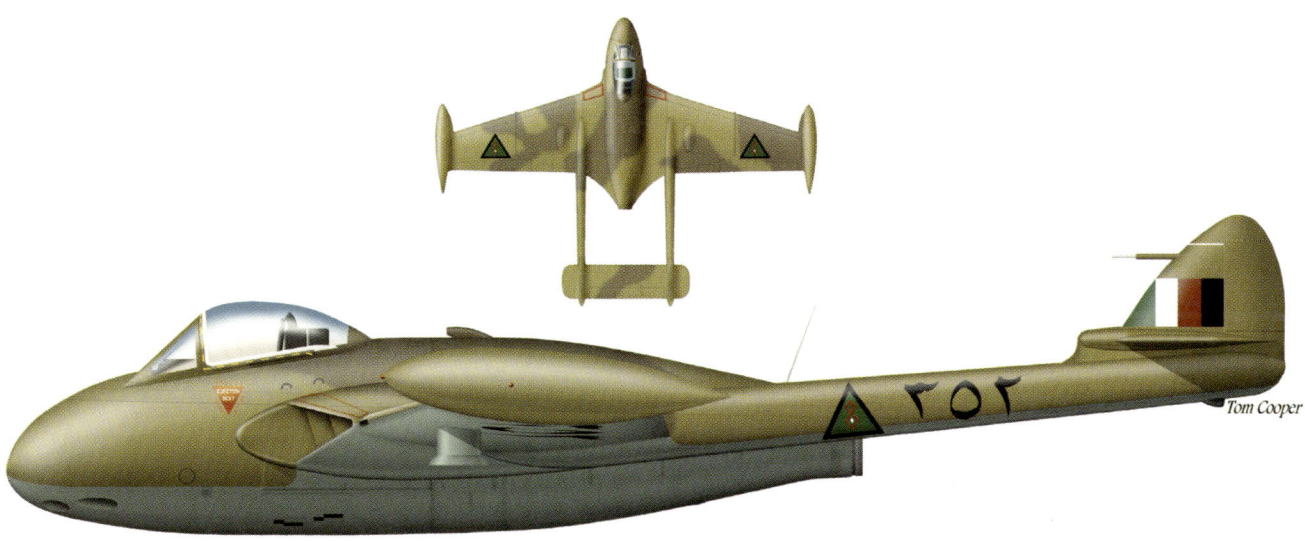

Iraqi Venoms wore a camouflage pattern similar to that of the Vampires. Even serial numbers were applied on lower wing surfaces in the same fashion. Shown is Venom FB.Mk.1 #352.

Meanwhile, the IrAF ordered 19 more powerful deHavilland Venom F.Mk.1 fighter bombers. Rebuilt into FB.Mk.50 fighter-bomber variant before delivery, the first of these arrived in August 1953, while the others followed at a monthly rate of between one and two aircraft. Venoms entered service with re-established No.6 Squadron, also based at Habbaniyah AB.

In 1956 the IrAF re-considered the equipment of its two leading jet units. The next year, the No.5 Squadron handed over its surviving Vampires to the Flying College, then based at Rashid AB, where these were subsequently used for advanced training. In turn, this unit was re-equipped with Venoms from No.6 Squadron, while the later received Hunters.

According to reports of British personnel seconded to the Iraqi Air Force only six FB.Mk.52s remained operational by 1958. However, when on 14 July 1959, a parade was held at Rashid AB, the No.5 Squadron put up 12 aircraft on display, including at least three two-seaters.

In 1961, the first Iraqi Vampire, T.Mk.55 #333, was sent to the UK for overhaul and fitting of ejection seats. Due to the poor state of the fuselage, the British felt forced to match the fuselage of ex-RAF T.Mk.11 XH316 to its wings and tail section, in turn practically creating a completely new aircraft.

Swiftly surpassed by the more powerful Venoms, Iraqi Vampires were used as advanced trainers during most of their career with IrAF. No less but three future Commanders of Iraqi Air Force began their career as fast jet pilots on the type, including Nimma Dulaymee (CO IrAF 1974-1976), Hameed Sha'ban (CO IrAF 1976-1977, and again 1984-1990), and Mohammad Jassem (CO IrAF 1979-1984). At least four Vampires were still in service with the Flight College, when this was enlargened into a Flight Academy and moved to Shoaibah AB, in southern Iraq, in 1964.

The very first Vampire to reach Iraq was T.Mk.55 #333, seen here on pre-delivery flight.
(Tom Cooper Collection)

Another view of the same aircraft, showing the camouflage pattern and position of markings to advantage.
(David Nicolle Collection)

Vampire T.Mk.55 #387 as seen before delivery.
(Albert Grandolini Collection)

deHavilland Vampire & Venom

In 1959, the No.5 Squadron was re-equipped with Soviet-delivered MiG-17s, and Venoms handed over to No.1 Squadron. The later unit operated them until 1966, by when all the surviving airframes expired their flying hours. This squadron was subsequently disbanded.

In IrAF service, Vampires and Venoms were armed with four 20mm cannons. They could also carry up to two 1,000 (454kg) GP-bombs or eight unguided rockets.

During the parade on Rashid AB, on 14 July 1959, the No.5 Squadron proudly put on display most of its Vampires (second row from the right), including three T.Mk.55s and nine FB.Mk.52s. (Ahmad Sadik Collection)

Front view of an IrAF Venom. Sadly, the exact serial number, usually applied on the cover of front wheel bay, is unreadable. (Ahmad Sadik Collection)

Iraqi Fighters, 1953-2003

Three Vampire FB.MK.52s on a training flight, showing most of the details related to their camouflage patterns to advantage. Sadly, it is unclear in which way the front fuselage of these three planes was camouflaged. (Albert Grandolini Collection)

The first Iraqi Venom FB.Mk.1, #352, as seen pre-delivery. (Ahmad Sadik Collection)

Formation of IrAF Venoms during a training flight in the mid-1950s, including (from left to right): #360, #355, #361 (probably), and #354. (Albert Grandolini Collection)

Camouflage, Markings & Serial Numbers

All Vampires and Venoms delivered to Iraq were camouflaged in dark earth over mid stone on upper surfaces. Two-seat Vampires should have had their lower surfaces painted in light mediterranean blue, while single-seat Vampires and all Venoms have had their lower surfaces painted in Sky Type S colour.

National markings were applied in six positions, and a fin-flash applied on the fins. No unit insignia is known to have been worn.

Taken during another parade on Rashid AB, the photo shows a row of No.5 Squadron's Venoms. #370 — the last delivered to Iraq — is in the front.
(Ahmad Sadik Collection)

Iraqi Fighters, 1953-2003

Table 2: Serial Numbers
Serial numbers were applied in black on the booms, and repeated below both wings: in flight direction below the starboard wing, and pointing towards rear on the port wings. Known serial numbers are:

IrAF Serial No.	Type	Delivery	Remarks
333	Vampire T.Mk.55	May 1953	
334	Vampire T.Mk.55	May 1953	
335	Vampire T.Mk.55	May 1953	
336	Vampire FB.Mk.52	May 1953	
337	Vampire FB.Mk.52	May 1953	
338	Vampire FB.Mk.52	May 1953	
339	Vampire FB.Mk.52	May 1953	
340	Vampire FB.Mk.52	May 1953	
341	Vampire FB.Mk.52	May 1953	
342	Vampire FB.Mk.52	August 1953	
343	Vampire FB.Mk.52	August 1953	
344	Vampire FB.Mk.52	August 1953	
352	Venom FB.Mk.1 (FB.Mk.50)	August 1953	
353	Venom FB.Mk.1		
354	Venom FB.Mk.1		
355	Venom FB.Mk.1		
356	Venom FB.Mk.1		
357	Venom FB.Mk.1		
358	Venom FB.Mk.1		
359	Venom FB.Mk.1		
360	Venom FB.Mk.1		
361	Venom FB.Mk.1		
362	Venom FB.Mk.1		
363	Venom FB.Mk.1		
364	Venom FB.Mk.1		
365	Venom FB.Mk.1		
366	Venom FB.Mk.1		
367	Venom FB.Mk.1		
368	Venom FB.Mk.1		
369	Venom FB.Mk.1		
370	Venom FB.Mk.1		
386	Vampire T.Mk.55		
387	Vampire T.Mk.55		
388	Vampire T.Mk.55		
389	Vampire T.Mk.55		
390	Vampire T.Mk.55		
391	Vampire T.Mk.55		

Chapter 2

Hawker Hunter

Hunter F.Mk.6/F.Mk.59A/F.Mk.59B & T.Mk.69

Service History

While increasing oil revenues permitted Iraq to obtain considerable numbers of combat aircraft at earlier times, the British were always reluctant to deliver the best they had at the time. The situation began to change in the mid-1950s. Iraq became a member of the so-called Middle East Central Treaty Organization, better known as "Baghdad Pact", in 1955, and when its government proved supportive for British interests during the Suez Crisis, in October 1956, London was quite fast with a reward. By the end of the same year, it has already donated six Hunter F.6 fighters to the IrAF. Within weeks, the USA joined this effort, funding ten additional Hunters through offshore procurement.

Taken during joint RAF-IrAF exercise in mid-1958, this photograph shows one of F.Mk.6s armed with 24 unguided rockets calibre 3in (78mm). This was standard air-to-ground attack configuration for the type at the time.
(David Nicolle Collection)

The only Hunter from original batch of 16 never to reach Iraq was #399. It crashed on 20 November 1956, in English Channel. Also illustrated is the insignia of No.6 Squadron.

F.Mk.6 #403 as seen wearing the fin flash in use on IrAF aircraft from 1958 until 1961. Also illustrated is the application of serial number on front undercarriage bay cover, and a 250gal. (1,000 litre) drop tank.

F.Mk.59B #694 belonged to the final batch of Hunters delivered to Iraq. It was flown as No.2 in a raid against Kfar Sirkin airfield, in Israel, on the morning of 5 June 1967. This artwork illustrates the weapons configuration selected for that raid: a bank for three 78mm rockets. Four such banks were carried under each wing, for a total of 24 rockets.

The RAF has set up a training program at the Hunter OCU in Chivenor, England, and a group of 20 Iraqi pilots arrived there already in October 1956. There they flew Hunter F.Mk.4s, before advancing to F.Mk.6s. These pilots were later to form the core of the re-organized No.6 Squadron, based at Habbaniyah AB, and under command of Major A-Razzaq.

Only five out of these first six planes reached Iraq: on 20 November 1956, 1st Lt. G. Mahmmoud experienced a stall during a training mission over the English Channel, and crashed. The pilot ejected safely and landed in cold water, but died later of hypothermia. The first Hunters were delivered to Iraq in April 1957; the other ten followed in December of the same year, and the No.6 Squadron was declared operational shortly later.

One of F.Mk.6s seen shortly after delivery to Iraq, in early 1957. (David Nicolle Collection)

Through 1958, the unit trained intensively, its pilots accumulating 120-150 hours on average, of which eight by night (or 12 a month, including two hours by night). Regular live firing exercises were staged, with pilots mainly training air-to-ground attacks. In March 1958, an air-to-air gunnery- and manoeuvring exercise was staged as well. It was during one of numerous training flights that the No.6 Squadron suffered its first loss in Iraq. On 3 October 1958, 1st Lt. Mufeed Saeed's aircraft suffered a catastrophic loss of hydraulic pressure, and crashed, killing the pilot. This was to become the last peace-time loss of IrAF Hunters in the first decade of their service.

IrAF Hunters were equipped mainly with standard avionics outfit and quite simple, but highly effective weapons. Like most of the other aircraft of this type in service elsewhere around the world, main armament consisted of four ADEN cannons, calibre 30mm, with 150 rounds each. In addition, banks of up to 24 unguided rockets – mainly calibre 78mm ("3in") – armed with warheads of various size, could be carried under the wings. British-made bombs calibre 500lbs and 1,000lbs (240kg and 454kg, respectively) were available as well, but seldom used. The F.Mk.6 version was powered by one Roll-Royce Avon 203 single-shaft turbojet, with 1,000lb thrust at full power, and has had extended-chord dog-tooth wings, for improved manoeuvrability.

Introduction of Hunter to IrAF was to prove the most decisive development, strongly influencing the history of this service for the following 30 years. Not only was the type to play an important role in the history of the entire country, or two wars with Israel, but also dozens of Iraqi Hunter-pilots rose high in rank, several becoming Commanders, or Deputy Commanders of the IrAF.

Barely half a year after becoming operational, Iraqi Hunters became the first ever to fire their weapons in anger. On the morning of 14 July 1958, a bloody coup by Iraqi armed forces swept the young King Feisal from power. In reaction, the pro-royalist elements of the 1 Iraqi Division, led by the veteran General Omar Ali, launched an advance on Baghdad from the south. In response, a Hunter from No.6 Squadron strafed their column while this was underway near Hillah, some 100km south of the Iraqi Capitol. The royalists stopped and then retreated: the republicans won the upper hand.

Following the coup, the new Government allied itself with the USSR, and the Soviets were very swift to provide considerable shipments of heavy weapons, including plenty of brand-new MiGs, as well as 120 instructors. The arrival of the later changed very little – if anything – in the IrAF. The unit structure, training and tactics, uniforms, and even traditions in most of the units remained very British-like. Nevertheless, the turbulent political scene in Iraq of the late 1950s and early 1960s has left its traces within the air force. Like the whole society was torn be-

Iraqi Fighters, 1953-2003

IrAF received four Hunter F. Mk.59s modified for carrying reconnaissance cameras in the nose, similar to FR.Mk.10 standard. One of these, serial number 660, was photographed shortly before delivery flight to Iraq.
(David Nicolle Collection)

Pre-delivery photograph of Hunter F.Mk.59A #570, taken at Chivenor, and showing details of national markings and serial number on the rear of the fuselage.
(David Nicolle Collection)

Hunter FR.Mk.10 #664 was the fourth example of this version delivered to Iraq. Note that the nose and the fin were painted in red or orange — the reason for this practice remaining unknown. All the other markings are rather standard in appearance.
(David Nicolle Collection)

Hawker Hunter

The first group of IrAF pilots with their British instructors, as seen during delivery ceremony at Chivenor, in January 1957.
(Ahmad Sadik Collection)

tween the Communists and Pan Arabists, Republicans and few remaining Royalists, this reflected strongly on the composition of the Iraqi military. The problem was that various political groups formed within various IrAF units too: No.6 Squadron was considered a "Pan-Arabist" unit. Correspondingly, a few of Iraqi officers including a number of Hunter-pilots – experienced diametrically opposed treatment by successive governments. Lieutenant-Colonel A-Razzaq, for example, became Base Commander of Habbaniyah, in July 1958, only to land in a jail, in March 1959. Barely five months later he was returned to service with the rank of Wing Commander, and posted to Habbaniyah again, to remain there until 1963.

It was no surprise, that the No.6 Squadron was directly involved in one of the following military coups. On 8 February 1963, when one of its pilots single-handedly mauled the No.9 Squadron – considered the "elite" of the IrAF, but manned almost entirely by "pro-Communist" pilots (see Chapter 4 MiG-19s). Other pilots from the same unit then flew a number of air strikes against the Iraqi Ministry of Defence, in downtown Baghdad, the headquarters of General Abdul Kareem Kassim. Appearing low over the Tigris River, divisions of four fighters would pop-up to rocket and strafe. Together with MiG-17s from No.7 Squadron, the No.6's Hunters flew over 60 sorties on the 8th and 9th of February alone.

These proved crucial in toppling the dictator. Indeed, shortly after this coup, Wing Commander A-Razzaq was appointed the new CO IrAF. Pleased with the performance of the Hunter, in 1964 the Iraqis ordered a second batch. The first to arrive were 22-ex-Belgian Hunter F.Mk.6s, upgraded to FGA.9-standard (10,150lb Avon 207 engine and heavier underwing load) before delivery, and designated F.Mk.59, F.Mk.59A and F.Mk.59B in Iraqi service. Recognizing the advantages of such a superb all-round combat aircraft, the Iraqis were swift to order another batch of 12 Hunters – all ex-Dutch F.Mk.6s brought to FGA.9 standard as well – already the following year. This contract was subsequently enlargened by 13 additional ex-Belgian aircraft, some of which were broken up for spares almost immediately on delivery. A total of up to six Avon 122-powered T.Mk.69 two-seaters were leased or ordered over the time, but it is unclear whether all were delivered. The last two Hunter F.Mk.59Bs arrived in Iraq only in May 1967.

Iraqi Fighters, 1953-2003

Hunter F.Mk.59A #579, as seen at Malta, during a refuelling stop on delivery flight, sometimes in 1964. (David Nicolle Collection)

Early Hunter F.Mk.59s entered service with No.29 Squadron, newly-established on 18 November 1965, with considerable help of personnel from No.6 Squadron, and based in Habbaniyah as well. Many F.Mk.59Bs, as well as four Hunter T.Mk.66s went to an Operational Conversion Unit, established at the same airfield, and manned entirely by British pilots, under command of Capt. Crow.

The increased number of aircraft enabled the IrAF to increase also the number of hours flown per pilot annually. In addition, the No.6 Squadron was deployed a number of times to Kirkuk and Mosul, in support of Army operations against rebellious Kurds, so that its pilots gained plenty of combat experience.

By 1967, the Habbaniyah Air Wing with its three Hunter units was considered the best outfit of entire IrAF, and it was no surprise when it became the first to receive an order to deploy closer to Israel. On 31 May 1967, in what was seen as a start of a deployment to Mafraq, in Jordan, No.6 Squadron was sent to H-3 airfield, in western Iraq. At the time, this was a small installation in the middle of desert, with very poor, rudimentary facilities. Few days later, its pilots were already briefed about "Operation Rashid", a series of commando- and air raids against various targets in Israel. This plan expected that Israel would open hostilities, but would get mauled while fighting Egypt, with the Arabs would then launching a joint counter-offensive.

On the morning of 5 June 1967, the No.6 Squadron was ready, and five Hunters attacked Petah Tikva airfield (former RAF Kfar Sirkin), near Tel Aviv. Armed with 24 British-made 78mm ("3in") rockets, this formation completed its mission almost flawlessly, claiming seven Noratlas and Dakota transports as destroyed on the ground, even if the last two Hunters were forced to break by one of Israeli MIM-23 HAWK-SAMs. Meanwhile, another formation of three Hunters raided Lod International Airport, though nothing is known about the results of this attack.

On the same afternoon, the Israelis returned the favour, by dispatching four Vautours escorted by two Mirages (Iraqis indicate there were four) to raid H-3. This attack achieved surprise and caught the Iraqis unaware, the Israelis claiming to have destroyed six MiG-21s, five Hunters, and a An-12 on the ground (according to other Israeli version, it was one An-12, six MiG-21s, two MiG-17s, and two

One of IrAF Hunter T.Mk.66s, together with pilot, as seen at Habbaniyah, in 1965. (Ahmad Sadik Collection)

Hunters. According to official IrAF records, the actual figures included three MiG-21s, one Hunter, one An-12, and a Dove light transport.

Though the Israelis claimed an almost complete destruction of Arab air forces on 5 June, on the following morning six Hunters were launched to attack Israeli armour near Jenin, in the Westbank. This time, the Israeli response came just at the time this formation was landing back at H-3, while two Hunters were taking off to provide top cover. One of the later crashed on take off, killing its pilot, but the wingman engaged a Mirage and a Vautour in a short but very confused air battle, only to get claimed as "shot down" by two different Israeli pilots.

Following this attack, the IrAF decided to evacuate all remaining fighters from H-3 to Habbaniyah, where the No.6 Squadron was reinforced by arrival of one Pakistani and one Jordanian pilot, during the following night. Still, the Israelis attacked for the third time on the late morning of 7 June 1967 – only to collide with a CAP of Hunters from Habbaniyah. In a pitched air battle, the Iraqis have lost one of their fighters, but in return the Pakistani pilot claimed two kills, his Jordanian wingman another one, and the surviving Iraqi pilot two other. According to surviving IrAF records, the Iraqis and Syrians later found four wrecks (two Mirages and two Vautours), captured two, and found the bodies of three Israelis.

Following the war, the Pakistani Flt.Lt. Saif-ul-Azam was decorated the Nowt al-Shuja'a (Medal of Bravery) for his achievements on 7 June. He went on to pursue a particularly successful career with the Pakistani and the Bangladeshi Air Force. His wingman, Fl.Lt. Ihsan Shurdom ultimately reached the rank of the

Iraqi Fighters, 1953-2003

Illustrating application of the serial number on the cover of front undercarriage bay, this photo shows the Hunter F.Mk.59B #628 as parked in a camouflaged blast pent at Habbaniyah AB, during the 1967 War.
(Ahmad Sadik Collection)

Chief of Staff Royal Jordanian Air Force. CO Habbaniyah Air Base from 1967, Wing Commander Sha'ban, became CO IrAF, in 1984; two of his Hunter-pilots from these times, 1st Lt. Hassan al-Khither and 1st Lt. Najdat al-Naqeeb, rose in rank to Director of Operations IrAF, and Deputy Commander IrAF, respectively. Two of their mounts, Hunter #575 (two confirmed kills) and #585 (one kill), were decorated appropriately, and slanted to end their days in the IrAF Museum, in downtown Baghdad. Before it was so far, however, they served with IrAF Display Team, and then – in the 1980s – with the Flight Leaders School.

Having lost only two aircraft during the 1967 War, the Hunters of the Habbaniyah Wing thus remained a potent force in the early 1970s as well.

In April 1973, 24 out of 28 operational F.Mk.59s and four T.Mk.69s, as well as 30 pilots from No.6 and No.29 Squadrons, were dispatched to Egypt, ostensibly for the purpose of joint exercises. In fact, Baghdad agreed with Cairo that these two units were to support Egypt in its fight for liberation of Sinai. Once in Egypt, IrAF Hunters were based at Qwaysina AB, in the Nile Delta, roughly half-way between Cairo and Alexandria, and only some 110km west of Suez Canal. Though under local command of Col. Natiq, IrAF, both units were officially directly reporting to the EAF High Command, which was granted permission to deploy them without any referring to the Government in Baghdad. The EAF further subordinated the No.6 Squadron as a close air support unit for Egyptian 2nd Army, and No.29 Squadron as a close air support unit for Egyptian 3rd Army. So it came that Iraqi Hunters were involved in 1973 War right from the start, beginning with the first

Hawker Hunter

A rare colour photograph of an IrAF Hunter F.Mk.59A, serial number #575, as seen during delivery flight, in March 1964.
(Tom Cooper Collection)

strike, flown at 14:00hrs on 6 October 1973, when they were mainly tasked with raiding Israeli SAM-sites and artillery positions. No Hunters were lost on these operations, but the Egyptians also cancelled the planned second wave, and Iraqis were rather disappointed.

On 7 October 1973, IrAF Hunters hit two Israeli HAWK SAM-sites again, using 40mm cannons only – with considerable success. No aircraft was lost during that mission. Later the same day, however, another Hunter was lost together with its pilot, near Ismailia.

8 October was to become the day of losses, especially for the No.6 Squadron. From a formation of four Hunters raiding Refidim AB, only Col. Natiq managed to reach Suez before bailing out, the other three planes being shot down over the airfield, and their pilots all killed. The No.29 Squadron, on the contrary, has spent the whole day in its ready-room, as the 3rd Egyptian Army never issued any requests for help.

On the morning of 9 October, two Hunters hit a massive column of Israeli vehicles in north western Sinai. The leader's plane was hit by ground fire, but both disengaged safely. It was only once they were back beyond the Suez Canal that the pilots encountered serious problem. When being taken under fire by Egyptian AAA: while the leader managed to limp back home, his wingman's aircraft was shot down and the pilot forced to eject. Dozens of other missions were completed during these initial days of the war without any particular problems and especially without losses. The matter of careful coordination of Iraqi operations with activities of the Egyptian air defences remained an issue, however, and the pilots had to pass the Suez Canal along carefully pre-selected corridors in order to avoid coming under friendly fire.

On 10 October, Qwaysina AB came under attack by the Israeli Air Force, which claimed destruction of six Hunters on the ground, although not a single Iraqi fighter was damaged. Several missions had to be postponed until the runway was repaired. The number of operational sorties flown by No.6 and No.29 Squadron decreased by 22 October 1973, when final missions of the war were flown, mainly because of an ever fewer requests from forward Egyptian Army elements.

Though Israeli narratives usually indicate the Iraqis to have avoided air combats with their fighters, and that four Hunters were shot down in clashes with Mirages or Neshers, former IrAF pilots indicate a completely different situation. They deny that any Hunters were shot down in air combats. They have often seen

Iraqi Fighters, 1953-2003

Hunter F.Mk.59A #570 as seen shortly after delivery (top) and then following the 1967 War with Israel (bottom). Of interest is application of red colour to nose and fin at later times, as well as kill markings on the right side below the cockpit, introduced as a commemorative measure. This was the plane flown by the (then) Pakistani Flt.Lt. Saif-ul-Azam when he shot down two Israeli fighters over H-3 airfield, on 7 June 1967. The aircraft remained in service in such disguise with the IrAF Display Team and then the Flight Leaders School IrAF, well into the 1980s.

Hunter #585 as the plane appeared in the early 1970s, while flown by an IrAF display team. It is shown with a nose to carry recce cameras and its entire tailfin – orange in colour. The Iraqis are known to have acquired four Hunters modified to FR.Mk.10 standard (including aircraft #660 through #664). Their noses were often mounted on other airframes as well. Also shown is the insignia of No.29 Squadron – the second IrAF unit to fly Hunters.

Hawker Hunter

Right-side view of the fins belonging to Hunters #585 (foreground) and #570, showing their former display colours to advantage. (Tom Cooper Collection)

Details of the kill marking commemorating the two kills scored by Pakistani Lt. Saif-ul-Azam, on 7 June 1967. The legend (in Arabic) reads: "Pilot Lieutenant Saif-ul-Azam". (Tom Cooper Collection)

Mirages patrolling high above, but there were no air-to-air battles what so ever. Sadly, no original IrAF documents are available to confirm or deny this.

In summary, the IrAF Hunter Wing in Egypt suffered a loss of eight Hunters and three pilots during this war. In turn it proved its worth despite facing a number of difficulties, including the slow pace of runway repairs at Qwaysina AB, and a lack of IFF-transponders compatible with Egyptian air defence system. The Iraqis were also trained to cooperate with forward observers on the ground, while the complex Egyptian chain of command took time to forward requests for any kind of missions, so that several times targets were not in place any more, when Iraqi fighters arrived in the area.

In the mid-1970s, the No.6 Squadron remained the sole unit to continue operating Hunters, as the No.29 Squadron was re-equipped with MiG-23BNs. By the 1980s, the surviving unit was disbanded and remaining airframes taken up by the Flight Leaders School – the Fighter Weapons School of the IrAF. They continued serving with this outfit, time and again flying few combat sorties against Iran (especially so in October 1980), until in the mid-1980s the last few Hunters were put into storage at Habbaniyah AB.

While flying Hunters, the No.6 Squadron became the only IrAF unit ever decorated with two medals: Iraqi Raidain Order (military version), the highest decoration for an Iraqi military unit, and the Egyptian Medal of Courage. The later was received by Maj. Izzat, on behalf of all the whole unit, from President Sadat, during latter's visit in Egypt, in 1974.

Details of #570's serial number on the right side of rear fuselage. The high-quality colours survived the last 40 years in actually a very good condition. (Tom Cooper Collection)

33

Iraqi Fighters, 1953-2003

Details of #585's fin, showing traces of the original fin flash below the one introduced in 1963. The rest of the original camouflage pattern and the serial number had previously been applied in white.
(Tom Cooper Collection)

A second look at the right side of #570's front fuselage reveals not only the remnants of the red colour that used to cover the entire nose section. A small yet important detail, two kill markings for aerial victories from the war with Israel, in 1967!
(Tom Cooper Collection)

Camouflage, Markings & Serial Numbers

All Hunters delivered to Iraq were camouflaged in dark sea grey and dark green over, with undersides in aluminium. Serial numbers were applied in black on rear fuselage on F.Mk.6s, and in white on the same place on all other variants. The serial number was always repeated in white on the cover of forward undercarriage bay. Roundel was applied in six spots, and all aircraft have got fin flashes as well.

Tired warriors: wrecks of three historic IrAF Hunters, as found at the former Habbaniyah AB, in western Iraq, in the early 2006. Fuselage of #575 is in the foreground, with the nose section of an unknown Hunter to the left, and that of #585 in the background. On the first view, there is nothing specific about these planes.
(Tom Cooper Collection)

Hawker Hunter

Table 3: Serial Numbers

IrAF Serial No.	Type	Delivery	Remarks
394	Hunter F.Mk.6	25 January 1957	ex RAF XJ677
395	Hunter F.Mk.6	10 January 1957	ex RAF XJ678
396	Hunter F.Mk.6	17 January 1957	ex RAF XJ679
397	Hunter F.Mk.6	24 January 1957	ex RAF XJ681
398	Hunter F.Mk.6	16 January 1957	ex RAF XJ682
399	Hunter F.Mk.6		Crashed on 20 November 1956
400	Hunter F.Mk.6	19 December 1957	ex XK143, RAF
401	Hunter F.Mk.6	19 December 1957	ex RAF XK144
402	Hunter F.Mk.6	19 December 1957	ex RAF XK145
403	Hunter F.Mk.6	19 December 1957	ex RAF XK146
404	Hunter F.Mk.6	19 December 1957	ex RAF XK147
405	Hunter F.Mk.6	19 December 1957	ex RAF XK152
406	Hunter F.Mk.6	19 December 1957	ex RAF XK153
407	Hunter F.Mk.6	19 December 1957	ex RAF XK154
408	Hunter F.Mk.6	19 December 1957	ex RAF XK155
409	Hunter F.Mk.6	19 December 1957	ex RAF XK156
567	Hunter T.Mk.66	Leased in 1963	ex Belgian AF IF19, sold to Hawker in 1967
567 (2nd)	Hunter T.Mk.66	Leased in 1963	ex Belgian AF IF67, sold to Hawker in 1967
567 (3rd)	Hunter T.Mk.69	24 February 1965	ex Belgian AF IF84
568	Hunter T.Mk.69	22 April 1964	ex Belgian AF IF68
569	Hunter T.Mk.69	14 May 1964	ex Belgian AF IF143
570	Hunter F.Mk.59	18 March 1964	ex Belgian AF IF6
571	Hunter F.Mk.59	18 March 1964	ex Belgian AF IF48
572	Hunter F.Mk.59	22 April 1964	ex Belgian AF IF14
573	Hunter F.Mk.59	22 April 1964	ex Belgian AF IF28
574	Hunter F.Mk.59	14 May 1964	ex Belgian AF IF21
575	Hunter F.Mk.59	14 May 1964	ex Belgian AF IF27, 2 kills in 1967
576	Hunter F.Mk.59	24 June 1964	ex Belgian AF IF140
577	Hunter F.Mk.59	24 June 1964	ex Belgian AF IF24
578	Hunter F.Mk.59	27 August 1964	ex Belgian AF IF88
579	Hunter F.Mk.59	27 August 1964	ex Belgian AF IF122
580	Hunter F.Mk.59	12 October 1964	ex Belgian AF IF07
581	Hunter F.Mk.59	12 October 1964	ex Belgian AF IF51
582	Hunter F.Mk.59	2 December 1964	ex Belgian AF IF142
583	Hunter F.Mk.59	26 November 1964	ex Belgian AF IF32
584	Hunter F.Mk.59	26 November 1964	ex Belgian AF IF0

IrAF Serial No.	Type	Delivery	Remarks
585	Hunter F.Mk.59	2 December 1964	ex Belgian AF IF126, 1 kill in 1967
586	Hunter F.Mk.59	17 December 1964	ex Belgian AF IF114
587	Hunter F.Mk.59	14 January 1965	ex Belgian AF IF75
626	Hunter F.Mk.59A	17 December 1964	ex Belgian AF IF97
627	Hunter T.Mk.69	14 January 1965	ex Belgian AF IF20
628	Hunter F.Mk.59A	14 January 1965	ex Belgian AF IF79
629	Hunter F.Mk.59A	24 February 1965	ex Belgian AF IF94
630	Hunter F.Mk.59A	13 April 1965	ex Belgian AF IF80
631	Hunter F.Mk.59A	13 April 1965	ex Belgian AF IF11
632	Hunter F.Mk.59	1965	ex KLu N-234
633	Hunter F.Mk.59	1965	ex KLu N-247
657	Hunter F.Mk.59A	1965	ex KLu N-253
658	Hunter F.Mk.59A	1965	ex KLu N-255
659	Hunter F.Mk.59A	10 January 1966	ex Belgian AF IF22
660	Hunter F.Mk.59A	10 January 1966	ex Belgian AF IF8
661	Hunter F.Mk.59A	16 May 1966	ex Belgian AF IF9, used as FR.Mk.10
662	Hunter F.Mk.59B	1965	ex KLu N-221
663	Hunter F.Mk.59B	1965	ex KLu N-205
664	Hunter F.Mk.59B	1965	ex KLu N-259
689 (ex-667)	Hunter F.Mk.59B	1965	ex KLu N-263
690 (ex-668)	Hunter F.Mk.59A	8 September 1966	ex Belgian AF IF71
691 (ex-669)	Hunter F.Mk.59A	13 October 1966	ex Belgian AF IF87
692 (ex-670)	Hunter F.Mk.59A	13 October 1966	ex Belgian AF IF54
693 (ex-671)	Hunter F.Mk.59A	14 November 1966	ex Belgian AF IF25
694 (ex-672)	Hunter F.Mk.59A	14 November 1966	ex Belgian AF IF135, No.2 in Kfar Sirkin raid
695 (ex-673)	Hunter F.Mk.59A	8 December 1966	ex Belgian AF IF74
696 (ex-674)	Hunter F.Mk.59A	8 December 1966	ex Belgian AF IF31
697 (ex-675)	Hunter F.Mk.59A	10 January 1967	ex Belgian AF IF59
698 (ex-676)	Hunter F.Mk.59A	10 January 1967	ex Belgian AF IF93
699 (ex-677)	Hunter F.Mk.59A	15 March 1967	ex Belgian AF IF99
700 (ex-678)	Hunter F.Mk.59A	15 March 1967	ex Belgian AF IF72, broken up for spares
701 (ex-679)	Hunter F.Mk.59A	13 May 1967	ex Belgian AF IF138, broken up for spares
702 (ex-680)	Hunter F.Mk.59B	13 May 1967	ex Belgian AF IF49, broken up for spares

Chapter 3

Mikoyan i Gurevich MiG-17

MiG-17F & MiG-17PF
(ASCC Code: "Fresco")

Service History

The arrival of first MiGs in Iraq followed a significant development in the history of the country. On 14 July 1958, King Feisal III and Crown Prince Abdullah of Iraq, together with the Iraqi Minister of Defence and a former Prime Minister, were assassinated in Baghdad during a coup d'état by elements of the Iraqi military, supporters of the pro-Soviet United Arab Republic (as Egypt was then officially designated). The bloody incident had repercussions for the IrAF as well: already on the same day, the prefix "Royal" was dropped from the designation of the air force. On 15 July, Brigadier-General J. Awqati, a staunch communist, assumed command of the IrAF. In a decision that might appear very strange for a former pilot trained in the Great Britain, almost immediately, he decided to change the source of aircraft acquired by Iraq, choosing the Soviet Union instead. His decision was most welcomed by Moscow. Within a few weeks, a big deal was signed with the USSR, which included agreements for delivery of MiG-17, MiG-19 fighters, Ilushin Il-28 bombers and Antonov An-12 transports. Several new squadrons were to be established by the IrAF to accommodate all the new aircraft.

The first Iraqi MiG-17F was #441, seen here shortly after delivery, in January 1959. Note the preliminary, post-1959-coup markings, with same colours like the old royal fin flash, albeit with a yellow disc in the middle. (Ahmad Sadik Collection)

Iraqi Fighters, 1953-2003

As originally delivered, Iraqi MiG-17Fs were painted in colour known locally as aluminium grey overall. Fin flash was applied on the top of the fin and consisted of the original RIrAF flag, with a yellow circle. Triangles were applied on the rear fuselage and undersides of wings only. Note that the serial number #441 in the case of the first IrAF MiG-17F, shown here was stencilled in black, apparently before delivery, but applied in various positions on front fuselage of different aircraft.

Within weeks of entering service, IrAF MiG-17Fs have gotten a new fin flash, introduced in 1959 and to remain in use as official insignia until 1963. Shown is MiG-17F #452.

An IrAF MiG-17PF, main mount of No.7 Squadron in the early 1960s. All the operational MiG-17PFs were sent to Syria, in October 1973, and their former pilots stress that they flew a number of CAPs and ground strikes during the war.

The influx of Soviet aircraft brought with itself another change as well. All British-made aircraft were delivered by the means of a ferry-flight from the UK to Iraq. Except for the biggest of their aircraft, such like bombers and transports, the Soviets were delivering all of their planes disassembled in crates, shipped from the port of Nikolayev, in Black Sea to the Iraqi port of Basrah. There the crates would be loaded on train and brought to Rashid AB. A group of Soviet technicians would assemble every aircraft. It would be test-flown by Soviet pilots before official delivery to IrAF. Until the aircraft was found operational and everything was in working condition, it officially remained Soviet property. The Soviets were responsible for any accidents occurring before the final handover, and they would pay compensation for any losses. This method of delivery was continued (with few exceptions) until the very last aircraft ever to be supplied to Iraq by the USSR, in 1989.

Though Soviet documentation indicates deliveries of 19 MiG-15bis in advance of MiG-17s, in early 1959, none were ever operated by the IrAF: most of the serials supposedly assigned to MiG-15s have been taken by Il-28 bombers.

Apparently, the other "MiG-15bis" sold to Iraq in 1959 were all MiG-15UTIs. Instead, Iraq did receive a total of 30 MiG-17s, in two batches (one of 10 and the other of 20 aircraft), during 1959. Reports according to which up to 100 MiG-17s (apparently including over 30 MiG-15s) have been supplied to Iraq cannot be confirmed. Nevertheless, it is easily possible that Iraq did receive more than 30 MiG-17Fs, but that a number of airframes were never assembled following their arrival. This appears to be supported by reports that some of these have been given to Egypt, together with a number of MiG-19s, and some other equipment, in 1963.

The equipment of MiG-17Fs as delivered to Iraq did not include any SRD-1M radar range finders, but it is possible that some aircraft were modified at a later stage. All Iraqi MiG-17s have got the Sirena 2 RWR (or "tail warning radar" as it used to be called at the time), the SRU-0 IFF transponders (with blade antennas installed in the fuselage spine), as well as EKSR-46 signal flares dispensers built into the right side of the fin (these were used for silent communication, not as defence from IR-homing missiles). The armament consisted of two NR-23 23mm and a single N-37 37mm cannons. Air-to-ground armament was apparently delivered at quite an early stage as well, and included D4-50 racks (carried instead of the drop tank) for bombs up to 250kg, or four unguided rockets. ORO-57K seven shot rocket pods were available as well. The IrAF selected an active duty unit as the first to transition to Soviet-made fighters: the No.5 Squadron, previously equipped with Venoms (now handed over to No.1 Squadron). The experienced staff of this unit guaranteed smooth conversion between two different aircraft, and indeed, the first group of Iraqi pilots converted to MiG-17s within only two months of the first examples of the type being delivered to Iraq, in January 1959. The whole process was completed without the use of MiG-15UTI two-seat conversion trainers. Two months later, on 13 February 1959, Major Khaled Sarah took over the command of No.5 Squadron.

Within a month of receiving their first mounts, the MiG-17Fs of No.5 Squadron already saw their first action. On 8 March 1959, there was a mutiny of the 5th Brigade Iraqi Army, based in Mosul, led by pan-Arabist officers who opposed the

Iraqi Fighters, 1953-2003

This is a reconstruction of the only IrAF MiG-17 lost during the October War, 1973, based on a photograph taken by David Nicolle in Qunaitara, in May 1974. The plane is shown wearing serial number 453, if this was really re-applied following arrival of No.7 Squadron's aircraft in Syria and application of camouflage, remains unknown. Also shown is top view, revealing details of camouflage, as well as UB-16-57 rocket launcher mounted on pylons as usually carried by Iraqi MiG-17s for fighter-bomber operations. A number of Iraqi -17s were flown by Syrian pilots during that war and vice versa.

Communists in the Ministry of Defence. On the following day, four MiG-17Fs from No.5 Squadron rocketed the makeshift radio station positioned in the 5th Brigade's HQs, using unguided rockets calibre 57mm.

This action was decisive in putting down the mutiny. As the number of available MiG-17Fs increased, the No.7 Squadron was also re-equipped with the type, in 1961. Since there was no two-seat training version of MiG-17 available, the following year the Soviets have delivered a batch of seven MiG-15UTI two-seaters. These have seen a short service with No.7 Squadron, before being turned over to the Flight College, and later the Air Academy, in order to facilitate transition of additional pilots. In the same year, also ten MiG-17PFs have been supplied. They entered service with No.7 Squadron, previously equipped with Hawker Furies. Following conversion training, this unit was based at Habbaniyah AB.

Next time IrAF MiG-17s were to see combat on 8 February 1963, when they joined Hunters from No.6 Squadron in a coup against the regime of General A. K. Qasim. Armed with unguided rockets and cannons, they were flown in a series of some 60 strikes against the building of the Ministry of Defence, in downtown Baghdad. The new government decided to disband the No.5 Squadron, on 20 December 1963, because most of the pilots from this unit were considered politically unreliable. All the aircraft and equipment were transferred to the No.7 Squadron.

For the rest of the 1960s, the later unit participated in a number of operations against rebellious Iraqi Kurds, in northern Iraq. At an unknown date in the 1960s, the No.7 Squadron formed an aerobatic team, equipped with four MiG-17Fs painted in yellow and black.

Mikoyan i Gurevich MiG-17

Though the Israelis fiercely deny any such claims, several Arab MiG-17-pilots have claimed an Israeli F-4E Phantom as shot down in air combat during the October War, 1973. This dramatic – but undated – photograph, taken on Golan at the time, and currently on display at the Military Museum in Damascus, shows a MiG-17PF pursuing an F-4E at a very low level. The then CO No.7 Squadron IrAF, Lt.Col. Shehab al-Qaisy, claimed an Israeli Phantom as shot down under almost exactly such circumstances, on 18 October 1973. The No.7 Squadron IrAF is known to have taken all of its PFs to Syria during that war.
(Photo by Oscar Ruf Wilson)

MiG-17F #452 with fin flash as in use from 1959 until 1963. (Tom Cooper Collection)

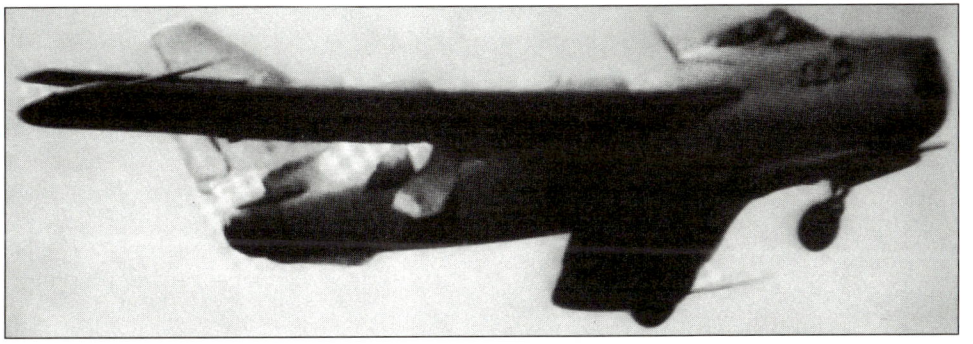

MiG-17F #445, as seen sometimes in the late 1960s. Note that the fin flash was moved down the fin, traces of triangles applied on undersides of both wings can be seen as well. The serial number is applied on a slightly different position than on #441, as seen above.
(Ahmad Sadik Collection)

Iraqi Fighters, 1953-2003

A unique shot of what was likely the only IrAF MiG-17 to be lost during the 1973 war. This photo was taken in May 1974 in the middle of Qunaitara, on the Golan Heights. The aircraft shows what appear to be a camouflage pattern then rather typical for Syrian MiG-17s, and obviously applied only after its arrival in that country (for reconstruction, see artwork on page 40). The fin flash is slightly larger than usually applied on Iraqi MiG-17s. (David Nicolle)

By October 1973, the No.7 Squadron was one of the largest IrAF units, operating no less but 34 aircraft, including MiG-17Fs and MiG-17PFs, as well as few remaining MiG-19s. At the time, the unit was based at Tammuz AB (also called al-Taqaddum), the then secondary facility next to the former RAF Habbaniyah, constructed in the 1940s. Following commencement of hostilities between Egypt, Syria and Israel, on 6 October, No.7 Squadron was put on alert. At noon of the following day, Lieutenant Colonel Shehab al-Qaisy, CO No.7 Squadron, led a formation of four MiG-17s to Syria. Transiting through H-3, by 8 October 1973, all the operational MiG-17s of the unit were based at al-Mezzeh AB, in western Damascus. During the October War with Israel, the pilots and aircraft of the unit were mainly tasked with flying CAPs over Syrian units on Golan, as well as raiding Israeli positions behind the frontlines. Iraqi MiG-17s clashed several times with Israeli fighters, and Lt.Col. al-Qaisy claimed one F-4E as shot down during a short air combat at low level, on 18 October 1973.

During the war, a number of IrAF pilots were seconded to various Syrian MiG-17-units, and vice versa. So it happened that when the only IrAF MiG-17 to have been lost in combat with Israelis was shot down over Qunaitara, in the final days of the war, it was flown by a Syrian pilot.

Following the October War, the No.7 Squadron returned to peacetime duty at Tammuz AB. With the end of Kurdish mutinies, and acquisition of more advanced

Mikoyan i Gurevich MiG-17

types by the IrAF, the career of the MiG-17 in Iraq came to an end in March 1975. The No.7 Squadron was disbanded, and several of its mounts put up as gate guardians in front of a number of IrAF bases.

Formation of No.7 Squadron's MiG-17Fs as seen over Baghdad, during a military parade in the late 1960s. (Ahmad Sadik Collection)

Camouflage, Markings & Serial Numbers

As originally delivered to Iraq, and for most of the 1960s, Iraqi MiG-17s were left in aluminium-grey all over. None were known to have been camouflaged before being sent to Syria, on 7 October 1973. During the October War, most of IrAF MiG-17s have been painted in sand yellow and dark olive green on the top portion. On the bottom portion on most of aircraft were painted in "Russian light blue", or left in aluminium-grey on others. The fin flash was initially carried on the top of the fin. In 1973 it was moved below the horizontal stabilizers. IrAF triangles were carried on the rear fuselage and undersides of wings only.

The aerobatic team established in the 1960s is known to have painted its aircraft in yellow and black. One of its former mounts is still on display in the "Green Zone" in Baghdad, as of today. No unit insignia is known to have been applied on any of IrAF MiG-17s.

43

Iraqi Fighters, 1953-2003

Table 4: Serial Numbers

Serial numbers of IrAF MiG-17s were applied in black on the front of the fuselage. Sadly, available photographic material is of rather poor quality and does not show their application on any other possible place, even though this was a standard practice on various other types in service during the 1960s and 1970s. The following serial numbers are mainly based on Soviet records, sadly, has proven as unreliable in several cases.

IrAF Serial No.	Type	Delivery	Remarks
440	MiG-17F	January 1959	
441	MiG-17F	January 1959	
442	MiG-17F	January 1959	
443	MiG-17F	January 1959	
444	MiG-17F	January 1959	
445	MiG-17F	January 1959	
446	MiG-17F	January 1959	
447	MiG-17F	January 1959	
448	MiG-17F	January 1959	
449	MiG-17F	January 1959	
450	MiG-17F	1959	
451	MiG-17F	1959	
452	MiG-17F	1959	
453	MiG-17F	1959	
454	MiG-17F	1959	
455	MiG-17F	1959	
456	MiG-17F	1959	
457	MiG-17F	1959	
458	MiG-17F	1959	
459	MiG-17F	1959	
460	MiG-17F	1959	
461	MiG-17F	1959	
462	MiG-17F	1959	
463	MiG-17F	1959	
464	MiG-17F	1959	
465	MiG-17F	1959	
466	MiG-17F	1959	
467	MiG-17F	1959	
468	MiG-17F	1959	
469	MiG-17F	1959	
508	MiG-17PF	1962	
509	MiG-17PF	1962	
510	MiG-17PF	1962	
511	MiG-17PF	1962	
512	MiG-17PF	1962	
513	MiG-17PF	1962	
514	MiG-17PF	1962	
515	MiG-17PF	1962	
516	MiG-17PF	1962	
517	MiG-17PF	1962	

Chapter 4

Mikoyan i Gurevich MiG-19

MiG-19S & MiG-19PM
(ASCC Code: Farmer)

Service History

The story of Iraqi MiG-19s is as short, as surprising and fascinating. In 1959 following acquisition of MiG-17 fighters and Il-28 bombers, the IrAF requested delivery of supersonic MiG-19S fighters. The Soviets agreed to deliver not only the basic version, but also MiG-19PMs equipped with RS-2U air-to-air missiles. At that time, one of the most advanced interceptor/missile combinations in their inventory. Originally, the pro-Communist regime in Baghdad envisioned equipping at least one unit, perhaps even the second, with the type. According to Soviet records, no less but 50 MiG-19s were indeed delivered. The IrAF, has certainly never taken as many, at best 14 aircraft were issued official Iraqi serial numbers, and the records show that there were never more than roughly a dozen of them in service. Apparently, the rest of the fleet reached Iraq, but was never officially handed over to the IrAF, but they remained packed in their crates. The 14 MiG-19s that certainly reached Iraq belonged to two variants. The "basic" MiG-19S was armed with three NR-30 cannons, and could also carry up to two 250kg bombs on BD-3-56 pylons (more often used for carriage of under-wing drop tanks). They could also carry ORO-57K rocket pods mounted on a removable pylon behind the main wheel doors. The other version was quite surprisingly, the MiG-19PMs, an interceptor

This very poor quality photograph is the only one available that clearly shows the details of markings as applied on Iraqi MiG-19s. The serial number 500 applied, in front of the cockpit. The fin flash (from the early 1963) is applied on the top of the fin. No national insignia is apparent on upper surface of starboard wing.
(Ahmad Sadik Collection)

45

Iraqi Fighters, 1953-2003

This reconstruction of the only available photograph of an IrAF MiG-19, sufficient to illustrate the details of its markings and serial numbers, shows the MiG-19S #500. Fin flash shown is as in use in the early 1960s, on the top of the fin. No triangles were applied on upper wing surfaces.

equipped with RP-2U radar and armed with four RS-2US air-to-air missiles. The exact break down between the two versions delivered remains unknown, but the number of PMs must have been limited, then their appearance remains hidden from the wider public, and can today only be traced in what is left of official IrAF records. Iraq thus became the only country outside the Warsaw Pact to have received this variant.

Interestingly, IrAF records indicate that its MiG-19PMs have also had two NR-30 30mm cannons – something unknown for any PMs so far. To facilitate a swift conversion, a few additional MiG-15UTIs have been delivered around the same time when MiG-19s and MiG-17PFs arrived, and also the first MiG-21F-13s were ordered, creating some "disorder" in the sequence of serial numbers.

Iraqi MiG-19 pilots and ground crews were carefully selected, then they were to man a unit destined for direct protection of the regime in Baghdad. As first, all were sent for training to what was then the City of Gorky (today Nizhni Novgorod), where MiG-19s were also manufactured until 1957. In the USSR, the Iraqi personnel has been instructed in Russian language, then received theoretical preparation in ground school. Afterwards, pilots underwent flight evaluations in the MiG-17Fs and MiG-15UTIS, before beginning with transition on MiG-19s. One of these planes crashed during the low level training on the night of the 3rd and 4th of June 1961, near Rashid airfield, claiming the only life of any Iraqi MiG-19 pilot ever.

Indeed, the conversion process was completed so swiftly, that Iraqi pilots began flying their MiG-19s in the country months before their unit was officially established. In fact, they even flew them in combat for the first time – flying strikes against Kurdish rebels in northern Iraq, in March 1961, before their squadron was completely organised. This new unit, based at Rashid AB, was the No.9, officially established on 11 June 1961, under command of Major Khaled Sarah. Without surprise, the No.9 Squadron was considered the elite of IrAF in the early 1960s. The unit would most likely continue to rate impressive operational and combat records. On 8 February 1963 its fate was sealed by a single stroke of the pen. Recognizing the importance of No.9 Squadron for the defence and power of the pro-Communist regime in Iraq, the coup plotters of 1963 decided that this unit had to be neutralized right at the start of their uprising, and almost simultaneously with an assassination of the then CO IrAF.

Mikoyan i Gurevich MiG-19

Supposedly showing ex-IrAF MiG-19S', this photograph was taken in Egypt, in the mid-1960s. Two aircraft can be seen wearing names "Basrah" and "al-Mousel" in white, in Arabic, below their cockpits. Sadly, these are not clearly readable, even on enlargened inserts. (David Nicolle Collection)

On the sunny and cool Friday morning, 8 February 1963, as the Indian instructors were preparing to make several training flights with their Iraqi students at the IrAF Flight College, then still stationed at Rashid AB, most of the No.9 Squadron's MiG-19s were lined-up on tarmac, refuelled and armed, ready to take off at short notice. Around 09:00hrs, all of a sudden a Hunter from No.6 Squadron, flown by 1st Lt. Munthir al-Windawi, appeared above: after circling the airfield at low altitude, al-Windawi popped-up, rolled and then dove over the line of MiGs, opening fire from all of his four ADEN 30mm cannons. Blasting all the way down, his attack turned the row of MiGs into a heap of smouldering wreckage: through the Indians later confirmed a complete destruction of only six MiG-19s, the No.9 Squadron was no more: with the loss of most of its PMs, it ceased to exist as a fighting force.

Following the coup, the remaining Iraqi MiG-19s never played an as important role as they used to play. The few MiG-19s' that remained operational (10 as of

47

January 1964, according to Soviet records) were assigned to the No.7 Squadron, with which they flew some additional strikes against Kurds, well into 1964. Then the IrAF decided to get rid of them, and – in agreement with the Soviets – the remaining airframes were sold to Egypt, following a decision to reform the No.9 Squadron as a MiG-21-unit. According to Soviet records, surviving Iraqi MiG-19s were packed in crates and re-distributed as follows:
- 18 to Afghanistan,
- 15 to North Korea, and
- 2 to Uganda.

It is likely that instead of being sent to Uganda (which was not known to have ever operated any MiG-19s) the two of disassembled Iraqi MiG-19s ended in Egypt as well. The reason for early retirement of MiG-19 from IrAF service has to do with at least as many technical, as political backgrounds. The type proved troublesome to fly, and even more to maintain, foremost because the engine as well as the hydraulic system tended to overheat. Though no MiG-19 ever crashed in Iraq, there was a series of incidents elsewhere, and, when additional MiG-21s and Hunters became available, the Iraqis deemed the type as surplus.

Although in theory no intact MiG-19 should thus have been left in Iraq after 1964, it is worth noticing that the full story about this type in IrAF service remains unknown until this very day. Namely, a reconnaissance picture of the Mosul AB, taken by an Iranian RF-4 Phantom, in 1980, shows an almost intact MiG-19 airframe on the local scrap yard. Also, a MiG-19 was put on display in the Museum of the Iraqi Air Force, in downtown Baghdad (the so-called "Green Zone"), and is there until this very day.

Camouflage, Markings & Serial Numbers

Originally delivered in aluminium-grey overall, for most of their service IrAF MiG-19s were painted light grey overall. No specific details about MiG-19PMs are known, but on MiG-19S' the fin flash from the period of early 1963 was applied on the top and, following the coup of 1963, the subsequent flash was worn further down the fin. Serial numbers were stencilled in black on the front of the fuselage. Their exact sequence remains unknown, but should have been #500 through #504, and #518 through #527. Although 50 MiG-19s should have been delivered to Iraq, only 14 of these entered service in 1961, and it appears that quite a number of MiG-17s and MiG-21s, perhaps even a Tu-16 or two, have taken up serial numbers originally slated for MiG-19s.

Egyptian sources indicate that after 1963 serial numbers were replaced by names of various Iraqi cities, applied in white below the cockpit. Such reports could not be independently confirmed by now.

Chapter 5

Mikoyan i Gurevich MiG-21

MiG-21F-13, MiG-21FL, MiG-21PFM, MiG-21MF and MiG-21bis
(ASCC Code: Fishbed)

Service History

Following negative experiences with MiG-19s, the IrAF was swift to order its first MiG-21 interceptors. A total of 12 MiG-21F-13s, and two MiG-21Us were delivered in 1962, packed in crates, from the Znamya Truda ("Banner of Labour") factory at Lukhovitsy, near Moscow: they all belonged to the second large batch of F-13s, manufactured in late 1961. They entered service with No.11 Squadron, officially established on 13 December 1962, and based at Rashid AB.

Powered by Tumansky R-11F-300 engines, Iraqi F-13s were equipped with SRD-5 Kvant ranging radar set, with detection range of 7km. Main armament included NR-30 30mm cannon, supplied with a short belt of only 30 rounds. Originally delivered without SRO-1 IFF, they were all equipped with SRO-2M Khrom-Nikel IFF once MiG-21FLs began arriving, in 1966. Other equipment included two underwing pylons for APU-13 launch rails (required to carry R-3S missiles, supplied together with MiGs), or UV-16-57 pods for unguided rockets calibre 57mm (with help of DZ-57 latches).

This very rare photograph of an early IrAF MiG-21F-13 shows #533, as it was in the mid-1960s. Of interest is the size of identification triangle on the rear fuselage, as well as the fact that no IFF-antennas can be seen anywhere around the aircraft: SRO-2M Khrom-Nikel IFF was added only once the MiG-21FLs were delivered as well, starting in 1964. Note the inclination of the fin flash — a standard practice by that time. (Ahmad Sadik Collection)

49

Iraqi Fighters, 1953-2003

Probably the most famous IrAF MiG-21 ever was this F-13, serial number 534, flown to Israel on 16 August 1966, by Capt. Monir Roufa. This action was organised by the Israeli foreign secret service, Mossad. The plane was extensively tested by the Israelis, providing their pilots with immense tactical advantages during the 1967 War, and after. Of interest is that the serial number has been repeated on the drop tank – in lieu of the well-known practice from a number of air forces equipped with MiGs and Sukhois.
(Tom Cooper Collection)

Like No.9 Squadron, the No.11 was mauled in attack of No.6 Squadron's Hunter against Rashid AB, on 11 November 1963, when five MiG-21s were destroyed on the ground. Contrary to No.9 Squadron, the No.11 has got an opportunity to hit back, on 18 November, when its MiGs attacked several strongholds of coup plotters, and destroyed them using UV-16-57 rocket pods.

Following that coup, the new Government in Baghdad first took care to reinforce its Hunter units, before returning to the USSR with orders for supersonic-, night- and all-weather interceptors, in 1965. This resulted in delivery of the first 16 MiG-21FLs (all built by Znamya Truda factory through the same year), which entered service with No.17 Squadron, an entirely new unit, officially established on 8 January 1966, at Rashid AB.

Powered by Tumansky R-11F-2S-300 engines, MiG-21FLs had a distinctive, much larger radome than F-13, that contained the first airborne intercept radar ever to be delivered to Iraq, R-2L. Mainly used to support the use of R-3S missiles, R-2L was also required for use of Kh-66 "beam-riding" guided missiles, 30 of which were delivered as well. Interestingly, the Iraqi pilots found them very complex and troublesome to employ. They were never used on MiG-21s. Instead, all were mounted on Mi-25 helicopter gunships and spent against Iranians, 17 years later!

No.11 and No.17 Squadron were operational at the time of 1967 War, and aircraft from both units were involved in several inconclusive air battles with Israeli Mirages while defending the H-3 airfield, in western Iraq, on 5 and 6 June 1967.

Through 1967, the first MiG-21PFMs (mounting R-21 radars, and a re-designed two-piece canopy) entered service with No.9 Squadron, but the unit did not become operational in time to participate in the war with Israel. PFM brought with them additional advanced avionics, including KAP-2 autopilot, a single-channel system stabilizing the aircraft on the roll axis only, and SRO-2 IFF (capable not only for answering IFF-requests, but also request answers from other transponders), with SOD-57M transponders. The later two worked in conjunction with RP-21 radar. Iraqi MiG-21PFMs were also equipped with GP-9 gun pods, housing a GSh-23 23mm twin barrel cannon, with 200 rounds – a weapon that the Iraqi pilots were

Mikoyan i Gurevich MiG-21

Several MiG-21F-13s remained in service long enough to get a coat of camouflage colours – including this example, found derelict at Habbaniyah AB, in early 2006. The plane shows what was apparently a standard camouflage pattern for this version in IrAF service from around the mid-1970s. Sadly, no serial number is visible. Note the lack of identification triangles on upper wing surfaces, but also the traditional inclination of the fin flash, as well as the appearance of IFF- and RWR-antennas around the fin and wings. Most of original Iraqi MiG-21F-13s ended as gate guards around various IrAF air bases.
(Tom Cooper Collection)

comfortable with, and which could be mounted under the centreline. But, their main weaponry still consisted of only two R-3S missiles.

By the end of 1968, a total of 36 MiG-21PFMs entered service with IrAF, equipping the re-activated No.9, and re-equipping No.11 Squadron, after surviving MiG-21F-13s were put into storage. The later unit was, time and again, flying combat sorties against Kurdish insurgents, in northern Iraq, in the late 1960s.

Monitoring the rapid deployment of Imperial Iranian Air Force, which included orders for McDonnell Douglas F-4 Phantom fighter-bombers, in 1969, the IrAF launched a big expansion and modernization plan, which included procurement of most advanced interceptors available. Following a careful examination of its range, speed, armament, versatility and advanced equipment, the IrAF decided to purchase specifically MiG-21MFs as replacement for all three previously available versions. Following a corresponding order, No.9 and No.11 Squadron were re-equipped with MiG-21MFs, by the end of 1971, and their previous mounts put into storage, or handed over to No.17 Squadron, which became an Operational Conversion Unit, with a considerable number of MiG-21Us in service. All three units were deployed to Syria, during the next war with Israel, in 1973; No.9 and No.11 Squadron flying MiG-21MFs, while No.17 Squadron brought with it several MiG-21PFMs as well. Their pilots claimed kills against five Israeli fighters, in exchange for four MiG-21s.

Iraqi MiG-21MFs were powered by improved Tumansky R-13-300 engines, and carried not only a new and more powerful RP-22 radar, but also an internal GSh-23L cannon. They were also equipped with four underwing pylons, and initially armed with disastrous R-3S missiles. In 1974, Iraq received the first batch of much improved R-13M1 missiles, but these have never entirely replaced the older variant in service.

Immediately following the war of 1973, the IrAF received another batch of 12 MiG-21MFs but also four MiG-21Rs (built at Factory No.21, in Gorky), a recon-

Iraqi Fighters, 1953-2003

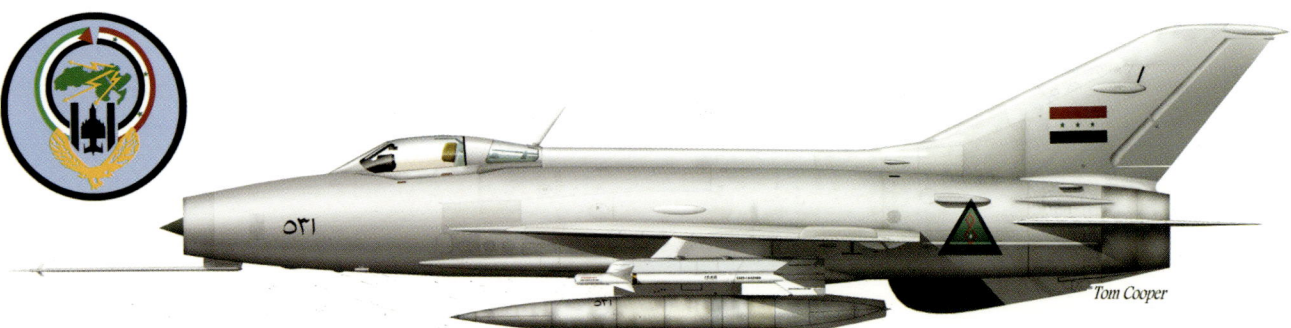

In best traditions of the time, IrAF MiG-21F-13s were painted in highly polished silver-grey overall on delivery. Fin flash introduced in 1963 was applied on the lower part of the fin; triangles on the rear fuselage and wing undersides only, thus introducing a long-standing tradition of no application of national markings on upper wing surfaces. The serial number was stencilled in black, and usually on the drop tank. Early Iraqi MiG-21 have had no Soviet-made IFF-systems, as IrAF radar network was not compatible with them. SRO-2 Khrom-Nikel IFF was introduced only once the transition to Soviet equipment was complete, in the second half of the 1960s. Also shown is the insignia of the first Iraqi MiG-21-unit: No.11 Squadron, which included a map of Arab countires, Palestinian and Iraqi flags and IrAF's Golden Falcon.

Aside from being painted in silver-grey overall, Iraqi MiG-21FLs have had a matt black anti-glare panel in front of the cockpit, applied over the cover for avionics bay in the same fashion this was the case with Egyptian MiG-21FLs. Most have also had their intake lips painted in the same colour as the dielectric radar cone. The aircraft numbered "668" served with the No.17 Squadron at the time of 1967 War (note the insignia of that unit), and is known to have flown combat sorties from H-3 airfield.

Reconstruction of the MiG-21PFM as seen on photograph on opposite page. Painted aluminium-grey overall, the plane otherwise wore standard IrAF markings. The insignia of No.17 Squadron included a map of Arabic countries in green (indicating support for "Pan-Arabic" ideas), stylised Iraqi national flag with two MiG-21s, and IrAF's Golden Falcon.

Mikoyan i Gurevich MiG-21

MiG-21PFM #863, as seen in the late 1960s. Note the GP-9 gondola under the centerline of the fuselage, as well as a missile rail under the port wing. The fin flash appears to be of standard size, and applied in standard position. (Ahmad Sadik Collection)

naissance version that formed the basis for MiG-21MF. Their GSh-23 cannons were removed and replaced by a camera/ELINT pod, carried under the centreline of the plane. Radar Warning Receiver antennas for SPO-3 system were fitted to wingtips, providing 360° coverage, and the avionics included AP-155 two-channel autopilot, providing pitch and roll control. Together with newest MFs, Iraqi Rs entered service with newly-established No.70 Squadron, based at Rashid AB.

Following another large order from the USSR in 1979, Iraq received its first 16 MiG-21bis. These entered service with No.47 Squadron, based in Kirkuk. This version featured several refinements in weapons and avionics (including SAU-23ESN autopilot, integrated with RSBN navigation- and PRGM landing system, theoretically enabling the aircraft to be guided from the ground and landed automatically; RSBN and PRGM were supported by the Pion antenna, mounted below the intake and on the top of the fin), the highly successful R-25-300 engine, and Almaz-23 radar. Armed with internal GSh-23, it arrived together with R-60 missiles. The IrAF received also dual launcher rails for the later, so that though usually carrying three drop tanks on intercept missions, Iraqi MiG-21bis' could still be armed with four air-to-air missiles.

By the time the war with Iran started, in September 1980, the IrAF could thus reach back on following MiG-21-units:
- No.9 Squadron, flying MiG-21MFs from Firnas AB (near Mousel), with detachment at Abu Ubaida AB (near al-Kut)
- No.11 Squadron, flying MiG-21MFs from Rashid AB, with detachment at Wahda AB (near Basrah)
- No.17 OCU, flying MiG-21FLs and MiG-21Us from IrAF Academy (Tikrit)
- No.47 Squadron, flying MiG-21bis' from al-Hurrya AB (Kikruk)
- No.70 Squadron, flying 4 MiG-21Rs and 12 MiG-21MFs from Rashid AB, with detachment of two MiG-21Rs at Wahda AB

Iraqi MiG-21s of the time were armed with Soviet-made, IR-homing R-3S, R-13M-1 and R-60 missiles, but also with French-made Matra R.550 Magic Mk.Is, the first batch of which was delivered already in the early summer 1980, and swift-

Iraqi Fighters, 1953-2003

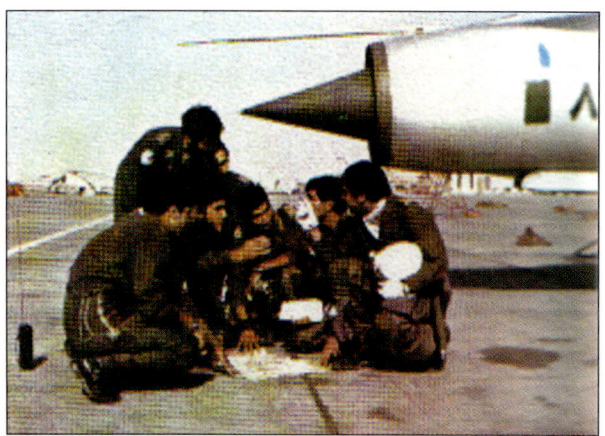

Group of IrAF MiG-21-pilots posing for camera in front of their MiG-21PFM. Of interest is the outline of intake which is in the same colour as on radome, and the first digit ("8") of the serial number can be seen.
(Ahmad Sadik Collection)

The few MiG-21PFMs that remained in service long enough, to have its serial number redone in 1988. Reconstruction of this MiG-21PFM can be seen at the top of page 56.
(Tom Cooper Collection)

ly integrated on a number of slightly upgraded airframes. Early during the war, MiG-21s formed the backbone of Iraqi air defence, and IrAF relied heavily on them for all sorts of related missions. Flying was extremely intensive, the IrAF keeping several CAPs airborne every day from early morning until the sunset. Initially, each CAP station was occupied by two MiG-21s, while later also single-ship patrols were flown. Aircraft from all five units became involved in scores of air battles along the border to Iran, and proved capable of successfully intercepting low-flying Iranian F-4s, with minimal support from ground control.

Immediately following the outbreak of war, relations between Baghdad and Moscow "froze". But, they improved in the early 1982, and IrAF immediately ordered 32 additional MiG-21bis, to replace surviving MiG-21MFs flown by No.9 and No.11 Squadrons. Indeed, by 1983, No.9, 11, 47 and 70 Squadrons were all flying MiG-21bis, while it remains unclear whether any were supplied to No.17 OCU as well, or if this unit was handed over the best remaining MiG-21MFs. By the time most of MiG-21MFs were badly worn out and in need of replacement (in fact, most of available MiG-21MFs, and available R-13 engines exceeded their service life by 1985). Although depending heavily on the type, the IrAF was not eager to expand its MiG-21 fleet, as it was considered obsolete due to its limited armament capabilities. While new aircraft of other types were purchased all the time, and MiG-23MF/ML, Su-22 as well as Su-25 fleets expanded during the war, MiG-21-units have had to soldier with what was at hand.

Surely enough, from 1983 the threat of Iranian raids against important installations inside Iraq decreased significantly, and tempo of operations slowed down. Soon enough IrAF MiG-21-units found themselves increasingly deployed for other tasks but air defence, including interception operations against Iranian helicopters. In 1986, a detachment of MiG-21bis' from No.9 Squadron was assigned to Wahda AB, with the task of operating against Iranian boats and small vessels roaming the northern Gulf. Painted in Dark Blue, Light Blue, and White, they were usually armed with four UV-32-57 rocket pods, and carried only the centreline drop tank. This detachment was disbanded in 1987, when the threat practically disappeared.

As of January 1991, the IrAF has had over 100 MiG-21s (including Chinese-made F-7s) at hand. Only six of these managed to launch into an operational sortie, and two were shot down when they turned into a flight of US Navy F/A-18 Hornets. Following the collapse of French-built KARI air defence system, the IrAF

Pilot mounting MiG-21MF #1190 (armed with UV-32-16 rocket pods), for a training mission, in the late 1970s. Of special interest is the very irregular pattern of green colour along the front fuselage, and IFF antennas under the intake. (Ahmad Sadik Collection)

stopped flying interceptor sorties, and with this the history of MiG-21 combat operations was over. Learning that medium-range air-to-air missiles are of paramount importance in modern air warfare, the Iraqis were forced to conclude that MiG-21s had no place in their front-line units.

Later on, during the 1990s, the IrAF reactivated some of its older MiG-21s – including few FLs and PFMs – mainly because of shortage of spare parts for other types. Most of these were captured almost intact in 2003, together with at least 25 MiG-21bis', and a similar number of two-seaters. The history of MiG-21 in Iraqi service thus came to an end after almost 40 years.

The F-7B Story

In 1982, the Iraqis were looking desperately for an advanced trainer, when reports surfaced about Egypt purchasing Shenyang F-7Bs from China. Decision was made to acquire 30 F-7B single- and FT-7 two-seaters as well. The aircraft were delivered to Giancalis AB, in Egypt, put together and test-flown by Chinese pilots, and then flown

One of several IrAF F-7Bs captured by US and Allied troops in spring 2003. Of interest is a washed out camouflage pattern, consisting of Sand, Dark Brown and Dark Olive Green, applied along a standard pattern pre-delivery. (Tom Cooper Collection)

Iraqi Fighters, 1953-2003

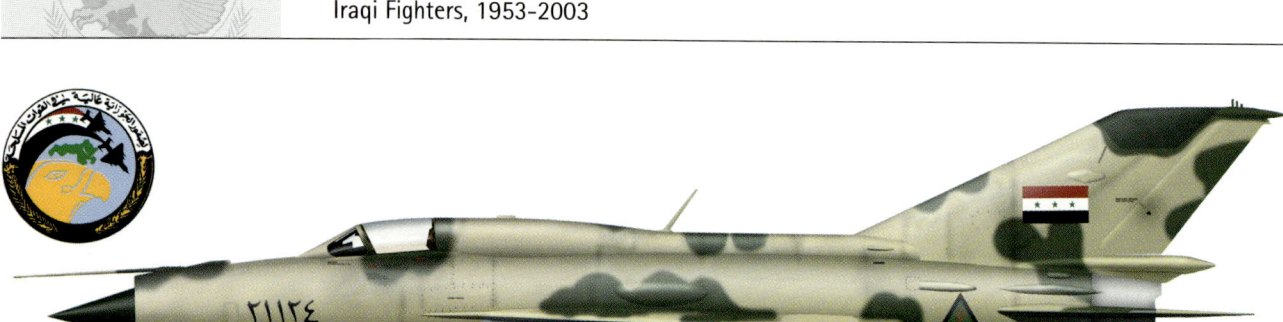

Like surviving MiG-21FLs before them, the remaining MiG-21PFMs have been camouflaged during the 1970s as well. This interesting example, serial number 21124 (found at the former Habbaniyah AB, in 2003), shows a camouflage pattern very similar to that – and applied in same colours – of early MiG-21MFs, as delivered by Znamya Truda factory. National markings are of standard size and in typical position.

No. 21112 was one of about 20 MiG-21FLs that survived long enough to be re-numbered following the war with Iran, in 1988. It was camouflaged in a modified pattern based on that of MiG-21MFs, but consisting of different colours. Camouflage colours from upper sides have been extended down the front edge of under-wing pylons – a widespread practice on Iraqi MiG-21s. The GP-9 gondola for GSh-23 cannon was left in aluminium-grey.

One of few confirmed pre-1988 serial numbers of IrAF MiG-21MFs was that of #1181. The plane is wearing a quite standard camouflage pattern for this version – reconstructed on basis of several photographs not useful for print. Of interest is that the serial number was replied at the top of the fin – a rather unusual practice for IrAF. The plane carrying an R-13M missile, as it was the practice during the early months of the war with Iran, in 1980. Also shown is the insignia of No.9 Squadron, depicting a MiG-21, powered by three arrows, and the Iraqi flag.

Mikoyan i Gurevich MiG-21

The pilot boarding the MiG-21MF (serial number 1184), is seen carrying the full load of four R-3S missiles, sometimes in the mid-1980s. Of interest is the irregular and very soft border between sand and light blue colours along the lower fuselage. (Ahmad Sadik Collection)

Another rare photograph of an early IrAF MiG-21bis, revealing details of camouflage pattern on upper surfaces. (Ahmad Sadik Collection)

to Iraq via Saudi Arabia. They never entered operational service, however: the IrAF considering them of little operational value. For most of the 1980s they were used for advanced- and weapons training. It was only in 1990 that, facing a confrontation with US-lead Gulf Coalition, the IrAF decided to prepare its F-7Bs for combat, as a back-up for MiG-21-fleet. The aircraft were armed with PL-2 missiles for this purpose (a Chinese copy of R-3S). Being considered no match of US- and British-made aircraft the opposition flew, they did not fly any operational sorties in 1991. A number of intact F-7Bs and FT-7s were captured by US and allied troops in 2003.

Camouflage, Markings & Serial Numbers

The original IrAF MiG-21F-13s were delivered in a highly polished overall natural metal finish, and have had gloss dark green air intake shock cones. They, as well as MiG-21FLs have also had the SOD antenna on the top of fin in natural metal. While the front half of the ventral fin on both versions was painted in gloss dark green as well, only the MiG-21PFMs became the first Iraqi 21s to have their SOD antenna painted in this colour as well.

The heat resistant parts of the exhaust cone have got a darker heat discoloured metal colour. All the FLs have also got a matt black "anti-glare panel" applied over the whole avionics bay cover, in the same manner this was done on most of Egyptian and quite a few early North Korean fighters of this version.

Iraqi Fighters, 1953-2003

This artwork is another match of a camouflage pattern seen on a photo too poor for print, and one of the few known original serial numbers of IrAF MiG-21MFs. The plane should have served with No.9 Squadron as of autumn 1980, nevertheless, and is known to have been one of few MiG-21MFs modified to carry French R.550 Magic air-to-air missiles.

Reconstruction of probably the most decorated IrAF MiG-21MF ever. During the war with Israel, in 1973, this plane shot down an Israeli Mirage, while flown by Maj. Namiq Saed-Al', over Golan. On 8 September 1980, Lt. Sadiq from No.11 Squadron flew this MiG to score the first kill against an Iranian F-4E Phantom. Both kills were scored by R-3S missiles. Of special interest is the unusual camouflage, worn also by MiG-21MF #1099, consisting of sand and dark olive green, seen on quite a few Egyptian MiG-21s in the 1980s. Sadly, its exact pattern remains unknown, and it is possible that there was an additional strip of olive green directly below or right behind the cockpit.

Probably the best-known Iraqi MiG-21, is also the most frequently misidentified one, this MiG-21R was captured by US troops at Ali Ibn Abu Talib AB, in 1991. Sadly, the only photographs of its right side are available, thus the camouflage pattern of the left side is a reconstruction of a MiG-21R. This particular aircraft is known to have served with No.70 Squadron IrAF in autumn 1980. Armament and equipment depicted here are as seen on the same photograph with four R-3S missiles and a D-21 reconnaissance pod.

Mikoyan i Gurevich MiG-21

A very rare photograph of an early IrAF MiG-21R (apparently has a serial number of 1828), showing the plane without any camouflage colours. Of interest is the carriage of drop tanks on outboard underwing pylons, a standard practice for all the Iraqi MiG-21 units. The recce pod under the centerline pylon appears to have been left in aluminium grey overall as well.
(Ahmad Sadik Collection)

For most of the 1960s, Iraqi MiG-21s were left in aluminium-grey overall. None are known to have been camouflaged before sometimes in the year 1972 or 1973. Photographs of early IrAF MiG-21MFs show even these left in this finish, at the times when all the MFs delivered to other countries were camouflaged. The then standard camouflage pattern for export examples for this version was introduced in the early 1970s, consisting of irregular splotches of olive drab on sand over, and light blue under, as seen on many early Egyptian MiG-21Ms and –MFs. Very few MiG-21MFs have got thick stripes of olive drab instead, colloquially described as "tiger stripes". All the MiG-21bis were subsequently delivered painted in similar pattern like most of MFs,

An IrAF MiG-21bis, apparently armed with R-60M missiles, seen on landing, sometimes in the early 1980s. Note the different pattern of green splotches along the front and mid-fuselage, as well as drop tanks on outboard under-wing pylons, left in bare metal overall.
(Farzad Bishop Collection)

59

Iraqi Fighters, 1953-2003

An interesting variant of an otherwise standardised camouflage pattern applied already before delivery. This IrAF MiG-21MF is wearing a serial number of an aircraft the wreckage of which was found at Habbaniyah, in April 2003.

Based on a photograph of an IrAF MiG-21bis captured by US troops at Ali Ibn Abu Talib AB, in 1991, this artwork depicts one of several Iraqi —21s known to have worn kill markings — probably commemorating an aerial victory from the early stages of the war with Iran. The plane should have served with No.11 Squadron at the time, and is shown carrying an R-13M missile on inboard underwing pylon: outboard underwing pylons were usually reserved for additional drop tanks. This plane was destroyed by US troops before these vacated Ali Ibn Abu Talib AB.

One of six IrAF MiG-21bis' that were sent for overhaul to Zmaj Works, near Zagreb, then in Yugoslavia, in 1990. With Yugoslavia falling apart, and Zagreb became the capital of Croatia, retreating Serbian military took these MiGs to Batajnica AB, near Belgrade. Most of them were destroyed by NATO air attacks, in 1999. Of interest is the camouflage pattern in sand and dark olive green colours, only reminiscent to pattern worn previously. No IrAF insignia is known to have been applied on any of these aircraft. Though some including #21204 depicted here, have got a set of Yugoslav Air Force insignia for transfer to Serbia, and this was later painted over in dark green.

60

Mikoyan i Gurevich MiG-21

Reconstruction of MiG-21bis as seen on p.64. Of interest is that the sand colour is slightly darker than the one used on MFs, as well as that green splotches grew in size. The plane is shown carrying another air-to-air missile frequently deployed by the IrAF already since the early days of the war: R-60M. All national markings were otherwise of standard dimensions, and applied in usual places. Note that the plane was carrying only the front Pion antenna: none can be seen carried on the top of the fin on any Iraqi MiG-21bis.

In early 1986, four MiG-21bis from No.9 Squadron, then based at Jalibah AB, were forward deployed to Wahda AB and painted in "naval camouflage", consisting of dark blue, mid-blue and white colours. Equipped with four UV-32-57 pods, these aircraft were exclusively deployed for attacks on hundreds of Iranian boats operating along the shores of northern Gulf and in Shatt al-Arab during that year. One of these MiGs is known to have been shot down during the Operation Karabla 3, in September 1986. This detachment was disbanded later in the same year, and three surviving planes repainted in their original colours.

The IrAF F-7B were never used in combat, and served for most of their career as advanced trainers. Nevertheless, they were prepared for combat in 1990/1991, and assigned to several MiG-21-units, and thus belong to be mentioned here as well. All the F-7Bs were painted in the same camouflage pattern, consisting of sand, dark brown and dark olive green, as shown here. Known serials are mentioned in Table 5.

61

Iraqi Fighters, 1953-2003

An IrAF MiG-21bis, probably from the No.9 Squadron, as seen in the early 1980s. Of special interest is the "classic" example for camouflage pattern applied on aircraft of this version in Znamya Truda factory. A number of MiG-21bis' seen on Iranian reconnaissance photos of various Iraqi airfields were painted in a very similar manner.
(Ahmad Sadik Collection)

Another IrAF MiG-21bis, this time carrying the full load of four air-to-air missiles (R-13M on inboard- and R-3S on outboard underwing pylons), and a drop tank under the centreline. Of interest is the Pion-antenna under the intake, as well as rather "spotty" appearance of the camouflage pattern.
(Ahmad Sadik Collection)

Two MiG-21bis (#21195 left and #21262 right), as captured by US troops in 2003, near al-Assad AB. Both wear a camouflage pattern typical for how it was applied already at Znamya Truda factory, before delivery. (via Tom Cooper)

A picture of MiG-21bis #21158, as found at Habbaniyah AB, in early 2006. Of interest is the camouflage pattern rather resembling that of Iraqi MFs, than any of other known "bis" airframes, as well as a different shade of green colour.
(Tom Cooper Collection)

Mikoyan i Gurevich MiG-21

Full view of the same aircraft. Both, the green and sand colours have obviously been washed out by sun and rain, and enough has been left to reconstruct large pieces of the pattern, as well as their original shades. It is possible that this was applied during an overhaul in some other country but USSR: the IrAF is known to have signed corresponding contracts with Czechoslovakia and Yugoslavia, in the late 1980s.
(Tom Cooper Collection)

albeit with a darker sand colour and much bigger splotches of green. A similar pattern, albeit with even bigger and rougher splotches of green colour, was applied on all MiG-21 bis overhauled in Yugoslavia, in the late 1980s. Surviving FLs and PFMs have got patterns similar to that of early MFs, albeit with an almost orange version of sand. Missile rails were initially left in natural metal overall, but later usually painted over either in light blue, or aluminium grey. Leading edges of underwing pylons on camouflaged aircraft were usually painted either in sand or olive drab. Centreline drop tanks were usually left in natural metal overall, but later on an ever increasing number of these were camouflaged in sand as well, sometimes with some splotches of olive drab. Reconnaissance pods for MiG-21Rs were initially left in natural metal, until sprayed over in light grey, probably at the time of the war with Iran. All camouflage colours applied on Iraqi MiG-21s were originally semi-gloss.

The air intake shock cone, forward half of the ventral fin and the SOD transponder antenna panel were painted gloss dark green.

In the 1990s, one of older MiG-21PFMs (serial number 1111) was taken out of storage for testing of radar-absorbent colour, and has had the port side of the fuselage and fin, as well as the whole port wing, painted in flat black overall. Though proving very successful, the IrAF lacked the time and means to pursue this project.

Fin flash was always carried on the lower centre of the fin. The situation with IrAF triangles was less unitary: most of the MiG-21s haven't got any on upper sides of wings, and only few exceptions are known.

At first three digit serial numbers were used. Later four, and finally five digit serial numbers were applied on front of the fuselage, in black. In very few cases, the serial number was repeated on the fin as well. Very few IrAF MiG-21s are known to have carried any special markings, and only few were seen with kill markings from the wars with Israel in 1973, and the war with Iran. All the maintenance, servicing and, emergency inscriptions were in Russian, and usually applied in black or white, sometimes in red or yellow. Iraqi F-7Bs were all painted in the same disruptive camouflage pattern, consisting of sand, dark brown and dark olive green on top, and light blue on lower surfaces. The FT-7s were painted in same colours but different pattern. They wore standard national colours, fin flash, and national insignia in six positions (including those on upper wing surfaces). Serial numbers were in black, applied on front fuselage.

Front left side of the same aircraft. Note the traces of the fresh sand colour with which the old serial number was painted over in 1988.
(Tom Cooper Collection)

63

Iraqi Fighters, 1953-2003

Aside from wearing the standard camouflage pattern, and national insignia of typical size and position, MiG-21bis #21240 clearly shows the Pion-antenna under the intake, as well as IFF- and RWR-antennas on the fin and wings.
(Tom Cooper Collection)

Mikoyan i Gurevich MiG-21

Serial number 21231 is one of very few Iraqi MiG-21bis' to have had a Pion-antenna on the top of the fin as well. The aircraft otherwise wore a camouflage pattern typical for this version. Of interest is the fin flash — which was not at an angle or inclined in this case, indicating it was probably applied already before delivery.
(Tom Cooper Collection)

Heavily retouched, but highly interesting photograph of an IrAF MiG-21bis, returning from a CAP in the early stages of the war with Iran, revealing many details of the camouflage pattern on upper surfaces. Of interest is also the application of identification triangle on the port wing, as well as the 800-litre drop tank carried on the centerline pylon, with upper surfaces apparently painted in the colour sand.
(Ahmad Sadik Collection)

Two MiG-21bis of No.9 Squadron, as seen on the afternoon of 25 September 1980 — from an Iranian RF-5A reconnaissance fighter. The camouflage pattern is quite clearly to be seen — so far that the plane to the left apparently carried an identification triangle on the upper surface of the left wing. A very rare appearance between IrAF MiG-21s. Note that both MiGs have also drop tanks mounted on their outboard under-wing hardpoints.
(Farzad Bishop Collection)

Iraqi Fighters, 1953-2003

A well-known photograph of MiG-21R, captured and subsequently destroyed by US troops at Ali Ibn Abu Talib AB, in 1991. Of interest is the camouflage pattern mimicking that of MiG-21MFs, but also including several differences — foremost in the area of mid-fuselage.
(Tom Cooper Collection)

Table 5: Serial Numbers

Serial numbers of Iraqi MiG-21s were applied in black on the front fuselage, usually in the form of a stencil at earlier times, but later almost exclusively with brush (especially so on camouflaged examples). In order to complete the picture, known serial numbers of IrAF MiG-21U/UM two-seaters and Chinese-built F-7B fighters were added as well, although these were never deployed in combat.

IrAF Serial No.	Type	Delivery	Remarks
Pre-1988 Serial Numbers			
521	MiG-21F-13	1962	Captured 2005, Qadessiya
522	MiG-21F-13	1962	Captured 2003, Qadessiya
528	MiG-21F-13	1962	Served with No.11 Sqn, fate unknown
533	MiG-21F-13	1962	Served with No.11 Sqn, fate unknown
534	MiG-21F-13	1962	Served with No.11 Sqn, flown to Israel
665	MiG-21FL	1966	Served with No.17 Sqn (possibly a MiG-21U)
666	MiG-21FL	1966	
667	MiG-21FL	1966	
668	MiG-21FL	1966	
669	MiG-21FL	1966	
670	MiG-21FL	1966	
671	MiG-21FL	1966	
672	MiG-21FL	1966	
673	MiG-21FL	1966	
674	MiG-21FL	1966	
675	MiG-21FL	1966	
676	MiG-21FL	1966	
677	MiG-21FL	1966	
678	MiG-21FL	1966	
679	MiG-21FL	1966	
680	MiG-21FL	1966	
681	MiG-21FL	1966	
682	MiG-21FL	1966	
683	MiG-21FL	1966	
684	MiG-21FL	1966	
702	MiG-21PFM	1967	Served with No.9 Sqn
703	MiG-21PFM	1967	
704	MiG-21PFM	1967	
705	MiG-21PFM	1967	
706	MiG-21PFM	1967	
707	MiG-21PFM	1967	
708	MiG-21PFM	1967	
709	MiG-21PFM	1967	
710	MiG-21PFM	1967	
711	MiG-21PFM	1967	

IrAF Serial No.	Type	Delivery	Remarks
712	MiG-21PFM	1967	
713	MiG-21PFM	1967	
714	MiG-21PFM	1967	
715	MiG-21PFM	1967	
716	MiG-21PFM	1967	
717	MiG-21PFM	1967	
718	MiG-21PFM	1967	
719	MiG-21PFM	1967	
720	MiG-21PFM	1967	
721	MiG-21PFM	1967	
722	MiG-21PFM	1967	
723	MiG-21PFM	1967	
724	MiG-21PFM	1967	
725	MiG-21PFM	1967	
726	MiG-21PFM	1967	
727	MiG-21PFM	1967	
728	MiG-21PFM	1967	
729	MiG-21PFM	1967	
730	MiG-21PFM	1967	
731	MiG-21PFM	1967	
732	MiG-21PFM	1967	
733	MiG-21PFM	1967	
734	MiG-21PFM	1967	
735	MiG-21PFM	1967	
736	MiG-21PFM	1967	
737	MiG-21PFM	1967	
818	MiG-21MF	1971	
819	MiG-21MF	1971	
820	MiG-21MF	1971	
821	MiG-21MF	1971	
822	MiG-21MF	1971	
823	MiG-21MF	1971	
824	MiG-21MF	1971	
825	MiG-21MF	1971	
826	MiG-21MF	1971	
827	MiG-21MF	1971	
828	MiG-21MF	1971	
829	MiG-21MF	1971	
830	MiG-21MF	1971	
831	MiG-21MF	1971	

Mikoyan i Gurevich MiG-21

IrAF Serial No.	Type	Delivery	Remarks
832	MiG-21MF	1971	
833	MiG-21MF	1971	
1019	MiG-21MF	1971	1 IDF/AF Mirage kill in 1973, 1 IRIAF F-4 in 1980
1051	MiG-21MF	1973	
1052	MiG-21MF	1973	
1099	MiG-21MF	1973	
1114	MiG-21FL		Post-1988 serial, RAM-coat testbed, captured 2003
1181	MiG-21MF		Served with No.9 Sqn
1184	MiG-21MF	1973	
1190	MiG-21MF	1973	
1195	MiG-21MF		
1232	MiG-21MF		
1598	F-7B	1983	Wreck in 2003
1828	MiG-21R		
4097	MiG-21bis		Captured Tammuz, 2003
6578	F-7B	1983	Captured 2003
Post-1988 Serial Numbers			
21023	MiG-21UM		Destroyed 1991
21027	MiG-21UM		Captured 2003, Qadessiya
21038	MiG-21UM		Captured 2003, Qadessiya
21040	MiG-21UM		Captured 2003, Qadessiya
21045	MiG-21UM		Captured 2003, Qadessiya
21054	MiG-21UM		Wreckage at Tammuz
21062	MiG-21UM		Captured 2003
21068	MiG-21UM		Impounded in former East Germany, in 1990
21069	MiG-21UM		Captured 2003
21071	FT-7	1983	Captured 2003
21072	MiG-21UM		Captured 2003
21073	MiG-21UM		Destroyed 1991
21079	FT-7	1983	Captured 2003, Qadessiya
21091	MiG-21UM		Captured 2003
21111	MiG-21FL		Captured 2003
21112	MiG-21FL		Captured 2003
21113	MiG-21PFM		Captured 2003, Firnas
21121	MiG-21PFM		Captured 2003, Tammuz
21124	MiG-21PFM		Captured 2003, Tammuz
21127	MiG-21PFM		Captured 2003
21137	MiG-21F-13	1962	Captured 2003, Tammuz
21138	MiG-21bis		Impounded in former Yugoslavia, 1990
21158	MiG-21MF		Captured 2003, Tammuz

IrAF Serial No.	Type	Delivery	Remarks
21168	MiG-21bis		Impounded in former Yugoslavia, 1990
21173	MiG-21bis		Captured 2003, Tammuz
21178	MiG-21MF		1 IRIAF kill, destroyed 1991
21182	MiG-21MF		Dumped at H-3, wreckage found in 2003
21186	MiG-21bis		Impounded in former Yugoslavia, 1990
21195	MiG-21bis		Captured 2003, Qadessiya
21196	MiG-21bis		Captured 2003, Qadessiya
21198	MiG-21bis		Impounded in former Yugoslavia, 1990
21202	MiG-21bis		Captured 2003, Qadessiya
21204	MiG-21bis		Impounded in former Yugoslavia, 1990
21206	MiG-21bis		Impounded in former Yugoslavia, 1990
21206	MiG-21bis		Wreckage 2003, Tammuz
21208	MiG-21bis		Captured 2003
21216	MiG-21bis		Captured 2003
21221	MiG-21bis		Captured 2003
21230	MiG-21bis		Captured 2003, c/n N75093445
21231	MiG-21bis		Captured 2003, Tammuz
21233	MiG-21bis		Captured 2003, Tammuz
21240	MiG-21bis		Captured 2003, Tammuz
21250	MiG-21bis		Captured 2005, Tammuz
21252	MiG-21bis		Captured 2003, Tammuz
21256	MiG-21bis		Captured 2003, c/n 766403361
21261	MiG-21bis		Captured 2003
21262	MiG-21bis		Captured 2003
21285	MiG-21bis		Impounded in former Yugoslavia, 1990
21302	MiG-21R		Destroyed 1991, A.I.A.Talib
21507	F-7B	1983	Captured 2003
21511	F-7B	1983	Captured 2003
22561	F-7B	1983	Captured 2003
22566	F-7B	1983	Captured 2003
22589	F-7B	1983	Captured 2003
22607	F-7B	1983	Captured 2003, ex 6578

Chapter 6

Mikoyan i Gurevich MiG-23

MiG-23MS, MiG-23MF, MiG-23ML & MiG-23BN (ASCC Code: Flogger)

Service History

The first 18 MiG-23MS interceptors and two MiG-23UBs airframes arrived in Iraq in 1974, after a group of selected MiG-21-pilots and technicians was trained at Lugovaya training centre, near Frunze (Buishkek), in the USSR. Already the training syllabus prepared for Iraqi pilots appeared quite strange: though none of the pilots has had less than 1.500 hours on fast jets, the Soviets treated them as cadets. Out of six months of training, more than four were spent on the ground: the actual flying time was minimal. Without surprise, pilots that hoped to be trained in flying the MiG-23 to its extremes, came back home with quite a sour feeling.

What followed once the planes arrived in Iraq, was a shock. The MiG-23MS proved to be a technological catastrophe, merely a structure around the powerful Tumansky R-29-300 engine, equipped with variable geometry wings and avionics of much-reduced standard, including Almaz-23 radar from MiG-21bis, with detection range of only 32km, and lock-on range of 18km at best – and that at medium or high levels. The Sirena III RWR was known to Iraqis as useless already from earlier encounters with the Israelis. The aircraft was armed with internal GSh-23L cannon (with 200 rounds), and first-generation R-3S missiles. Many Iraqi pilots have seen more advanced missiles while in Syria during the 1973 War

A photo of an IrAF MiG-23MS shot at an unknown date (probably in 1982), and captured from Iraqi archives by Iranians. Sadly, the serial number was deleted by Iraqi censors. It is one of few shots clearly showing the standard export camouflage pattern for MiG-23MS, as applied to Egyptian, Iraqi, Libyan, and Syrian aircraft.
(Farzad Bishop Collection)

71

Iraqi Fighters, 1953-2003

These two artworks represent a reconstruction of a heavily burned out MiG-23MS, found by US troops near al-Bakr AB, in 2003. The first artwork shows what was apparently its original serial number. There was a Mirage F.1EQ with serial number 4012 also while serving with No.39 Squadron, in the early 1980s. Sadly, no details of camouflage pattern on upper surfaces are recognizable. Of interest is the serial number repeated below the fin flash, but in different style, as well as maintenance stencilling, which was mainly in English. Modified to carry the R-13M-1 missile in 1976, this aircraft could carry two such missiles on underwing pylons, and two R-3S on pylons under the fuselage. The second artwork depicts the same aircraft, but in the state as found in 2003, including the insignia of No.59 Squadron on intake, and a post-1988 serial number, applied on a splotch of fresh sand colour below the cockpit.

Based on an out of focus still photo taken from a video showing the "personal" mount of CO No.39 Squadron as around early 1981. This artwork depicts one of 18 MiG-23MS originally delivered to Iraq. Camouflage consisted of Sand colours (approximately FS13523), Dark Earth (FS20095) and Green (FS30098). Undersurfaces were in Light Blue-Grey (FS35622), and applied according to standardised pattern for all export MiG-23s. Serial number remains unconfirmed but the plane is known to have worn unit insignia on left intake (only), like all the other aircraft of No.39 Squadron.

with Israel. Interestingly, in 1974, the Soviets began delivering more advanced R-13M-1 missiles together with MiG-21MFs, by that time. The reason why none were supplied together with MiG-23MS' thus became a matter of controversy.

Worst of all was that the Soviets did not teach the Iraqis – nor supply them with technical documentation - about aerodynamic limitations of the MiG-23. Former MiG-21-pilots, accustomed to an almost vice-free aircraft that could be flown under all conditions, were thus facing previously almost unheard of limitations when flying MiG-23s. The result was a series of accidents that significantly depleted the No.39 Squadron during the first few years of its existence.

Meanwhile, in 1976, the IrAF purchased the first batch of 18 MiG-23BN fighter-bombers. The aircraft entered service with re-established No.29 Squadron, a former Hunter-unit, stationed at Ali Ibn Abu Talib AB. The second batch of 18 MiG-23BNs entered service with newly-established No.49 Squadron, based at Abu Ubaida AB. Though as supplied to Iraq these aircraft have had the "Delta" guidance pod for Kh-23 missiles built in, no such weapons were supplied, and they were mainly armed with UV-16-57 and UV-32-57 rocket pods. General purpose bombs like FAB-100s, -250, and -500 (mainly of M54-series) were available as well, but deployed rather seldom. Avionics was again of poor standards: though equipped with the AKS-5 cine camera housings, and appropriate blisters in place, the first two batches have got no SPS-141 Siren ECM-gear. Nevertheless, they were some of the earliest examples equipped to carry 800-litre drop tanks under outside wing sections. Whenever these were carried, the wing could not be swept back. Tanks had to be jettisoned together with pylon before the aircraft went into action. Though such limitations, the IrAF is known to have made extensive use of drop tanks on its MiG-23BNs at least during the opening strike against Iranian airfields, on the afternoon of 22 September 1980.

The condition of Iraqi MiG-23-fleet improved only slowly, with pilots "learning by doing". No less but 12 aircraft of all three variants (including two-seat MiG-23UBs) were lost in flying accidents by 1978, when the IrAF decided to modify underwing pylons of remaining MiG-23MS' for carriage of two R-13Ms. Though all the losses were replaced by the Soviets, the feeling about the original versions delivered was never particularly positive.

In 1979, the No.39 Squadron was moved from Rashid to al-Wallid AB – the former H-3, in western Iraq. While there, the unit has had several non-lethal encounters with Israeli reconnaissance planes and even F-15 Eagle fighters, in the early 1980s.

Shortly before the war with Iran, the unit was moved again, this time to Tammuz AB, near Habbaniyah, with the task of providing air defence of the Iraqi Capitol. Without ground-based long-range early warning radars capable of detecting and tracking low-flying Iranian fighter-bombers in service at the time, the efficiency of Iraqi MiG-23MS' was poor. Pilots of No.39 Squadron were usually scrambled too late and when in the air, they were on their own to search for their opponents. Tammuz AB was quite a distance away from main battlefields. Therefore, during the autumn 1980 detachments of two MiG-23MS were distributed to several forward airfields, from where they began operating over the battlefields. By autumn 1981, the CO No.39 Squadron claimed five Iranian helicopters as shot down, mainly by GSh-23 cannon while flying out of Wahda AB.

Iraqi Fighters, 1953-2003

An Iraqi MiG-23MF known to have worn this serial number in the late 1980s. There was also an MF with a kill marking from the war with Iran, this artwork is actually a reconstruction based upon several poor photographs of various MFs. The camouflage pattern is still easy to recognize as standard for export aircraft, even if the form of the green splotch around the cockpit was rather unusual. The plane is illustrated carrying a R-23R and a R-60M missile.

Illustrating the standard appearance of IrAF MiG-23MLs as in the late 1980s and early 1990s, #23254 is shown carrying an R-24R missile under the wing, and an R-60MK under the fuselage. Note the same dark grey colour of radome and all other dielectric panels, as used on MiG-23MFs previously. All the markings are in standard positions, and it appears that at least some of Iraqi MLs have had national insignia on upper wing surfaces.

Iraqi MiG-23BNs wore essentially the same camouflage pattern like MiG-23MS', the main difference being a slightly less reddish shade of brown and green colours. Like on all IrAF MiG-23s, the fin flash was applied with inclination. The plane is shown with national markings applied on upper surface of port wing, though these were not applied on all Iraqi MiG-23s. Also illustrated is the insignia of No.29 Squadron. Note that the shadow of the aircraft was changed to that of a MiG-23 – compared to earlier times when a shadow of Hunter was applied instead, while the unit was flying that type. The plane is shown carrying a FAB-250ShN parachute-retarded bomb, known to have been used by No.29 Squadron for attacks on Iranian airfields, on the afternoon of 22 September 1980.

74

Meanwhile, MiG-23BNs of No.29 and No.49 Squadrons were busy supporting Iraqi ground forces on the southern battlefields, and flying interdiction strikes. Mainly operating at low levels, they were facing the full range of Iranian air defence arsenal, and losses were heavy. Twelve aircraft being lost within the first year of war.

Once the relations to the USSR improved, the No.39 Squadron was immediately re-equipped with 16 MiG-23MF interceptors, handing over their former mounts to newly-established No.59 OCU, based at Tammuz AB. The later unit took over the training of all future Iraqi MiG-23 pilots. MiG-23MFs were still powered by R-29-300 engines, but equipped with S-23E weapons suite, incorporating the Sapfir-23D radar, R-23R and R-60 missiles.

With MiG-23MF the Iraqis felt they have finally got the interceptor they expected to have already back in the mid-1970s. Its Sapfir radar had a relatively good detection range (at least when compared with MiG-21), and was more reliable. Most important of all, the type was equipped with medium-range, semi-active radar homing air-to-air missiles, R-23Rs, and much improved R-60s.

This recovered the sense of pride within No.39 Squadron considerably. Pilots started feeling superior even to their colleagues flying French-built Mirages. Still, the unit was not as successful as the first Mirage F.1EQ-outfit, and claimed only two confirmed kills during air battles in 1982 and 1983.

In 1983, the USSR delivered a batch of 18 new and improved MiG-23BNs. While externally closely resembling the MiG-27K, however, their avionics standard remained quite similar to that of the "basic" BN, even if they were finally equipped with SPS-141 ECM-systems. These 18 fighters were distributed between No.29 and No.49 Squadrons.

Less than a year later, the Soviets began delivering 16 much improved MiG-23ML interceptors to Iraq. These first entered service with No.73 Squadron, based at Ali Ibn Abu Talib AB. MiG-23MLs were equipped with N003E Sapfir-23ML radar and TP-23IRST (similar to TP-26 of MiG-25PD), and powered by R-35F-300 engines. This radar was actually the first Soviet system with a reason-

Pilot of No.39 Squadron proudly posing next to his mount, a MiG-23MS, following the 1976-upgrade during which R-13M-1s compatibility was added (note the missiles mounted on under-wing pylons). Of interest is the national insignia as applied on undersurfaces of both wings -. pointing forwards when wings are swept fully back.
(Ahmad Sadik Collection)

Another shot showing the bottom of an IrAF MiG-23MS: note the awkward position of identification triangles, resulting from the fact that they were applied pointing forwards when wings were swept fully forward. Also of interest is the standard armament configuration, consisting of two R-13MS and two R-3S.
(Ahmad Sadik Collection)

Iraqi Fighters, 1953-2003

A very rare shot of an early IrAF MiG-23BN, with its original serial number of 1428. Camouflage pattern is standard for export aircraft sold also to other Arab nations, and serial numbers are of representative size, usually applied on IrAF MiG-23s of the mid-1970s. (Ahmad Sadik Collection)

able look-down/shot-down capability out to a range of some 30km, as well as first with a dedicated dogfight mode. The system proved quite problematic to maintain and was suffering from frequent malfunctions, yet it offered also enhanced ECCM-capabilities when comparable with earlier Soviet radars. Main armament consisted of R-24R as well as R-60MK missiles (in addition to the standard GSh-23L internal mount), which proved very lethal early on. During their first air combat, on 11 August 1984, two MiG-23MLs from No.73 Squadron scored the first confirmed kill against an Iranian F-14A Tomcat interceptor. For the rest of the war, they have also scored confirmed kills against Iranian F-4E Phantoms, Fokker F27 transports, and even an Israeli UAV.

Second batch of 16 MiG-23MLs entered service with No.63 Squadron, based at the newly constructed al-Bakr AB, north of Baghdad, in 1985. According to Soviet sources, a total of some 55 MiG-23MLs were delivered to Iraq. Interestingly, Iraqi sources, supported by US intelligence reports made available to authors though FOIA, cite that 36 fighters of this version were still available to the IrAF as of January 1991.

In total, there were 63 operational MiG-23MLs, -MFs, and –MS' in Iraq at the time, and they were very active in the first few nights of the war. On the first night, MiG-23MLs from al-Bakr engaged at least two USAF F-111Fs, and fired several missiles at them, though no confirmed kills have been scored. Another ML has had a particularly interesting encounter with a Lockheed F-117A over Baghdad: though the Iraqis could not track the F-117 with their radars, they brought their fighter on a collision course with the "stealth" fighter, and the crew of a USAF E-3A Sentry AWACS was forced to order the pilot to drop to a lower level in order to avoid a collision.

During the war, almost a squadron's worth of MIG-23BNs were destroyed at various airfields (foremost at Ali Ibn Abu Talib AB), eight various MiG-23s were evacuated to Iran, and two were shot down in air-to-air combats. During the 1990s, the MiG-23ML became the major interceptor within the IrAF, and Iraqis invested heavily into improving the type with help of available resources and equipment. The type

Mikoyan i Gurevich MiG-23

was equipped with French-made Remora ECM-pods (carried on port underwing pylon), SPO-15 RWRs and, ASO-2 chaff & flare dispensers taken from stored Su-22M-4Ks. This configuration proved capable of jamming US fighters sufficiently to prevent Iraqi MiG-23MLs from being hit by quite a few AIM-120 AMRAAMs fired at them, especially in 1999. One MiG-23ML claimed as shot down by US fighters in early 1999, was equipped with Remora. The pilot did a mistake while attempting to land, and went for a second attempt when it run out of fuel. The pilot attempted to glide back to the runway, but crashed in the process. The IrAF is known to have suffered several losses of MiG-23s during such landing attempts between the 1970s and 1990s. Unfortunately, all the involved pilots perished.

Despite plenty of activity between 1999 and 2003, during the invasion of 2003, the Iraqi MiG-23-fleet remained grounded, as it was clear that it would have not the least chance to challenge the US aerial supremacy.

Though taken 12 years after each other, and showing two different IrAF MiG-23BNs (#23173, left & #23166, right), these photographs show that most of the fleet's aircraft have had the same camouflage pattern, applied in same colours. Of interest on the later plane is addition of ASO-2 chaff & flare dispenser. Number 23166 was also equipped with SPS-141 ECM-system.
(via Tom Cooper)

Camouflage, Markings & Serial Numbers

All MiG-23s delivered to Iraq wore a standard disruptive camouflage pattern consisting of colours at best described as follows: Yellow-Sand (BS381C/388) or Beige (FS13523), Dark Earth (BS381C/450 or approximately FS20095), and Green (FS34098) on upper surfaces, and Light Blue-Grey (FS35622) on all lower surfaces. The radome and all dielectric panels were painted Mid-Grey (FS26152). Though unrecognized by Western observers at the time, this was standardised for all export aircraft, and there were only very few deviations between aircraft delivered not only to Iraq, but also to Ethiopia, Libya and Syria. Even the examples returned from overhaul in the USSR usually kept this pattern. Serial numbers were applied in black on front fuselage, however, but only few serial numbers from pre-1988 period remain known. Several early MiG-23MS have had their radomes painted white. Apparently, this practice all but disappeared by the late 1970s, by when all the radomes were painted mid-grey, just like all the dielectric panels. Subsequent MiG-23MFs and MiG-23MLs have had their radomes and all dielectric panels painted in dark grey.

Iraqi Fighters, 1953-2003

Sadly, no photos from better times of IrAF MIG-23MFs are available. This one shows three derelict examples, found by US troops at al-Bakr AB, in 2003, already full of graffiti. Some details of camouflage pattern and serial numbers are recognizable, as well as a darker shade of sand colour, #23127 (centre) and #23124 (right). (US DoD)

This photograph shows the starboard side of MiG-23ML #23270 (left), as well as the sister-ship #23278 to advantage. The later plane was also equipped with SPO-15 and ASO-2 dispensers taken from Su-22s. (US DoD)

MiG-23ML #23270 at Balad shows beside camouflage pattern, national markings, and APU-24 launch rails for R-24 missiles less obvious details, such like SPO-15 Siren-3 RWR on the top of the fin and ASO-2 chaff & flare dispensers on the back of the fuselage. Both items have been taken from Su-22M-4Ks in the mid-1990s. (US DoD)

The few of rarely available good photographs of IrAF MiG-23s, are showing the ML #23255, while in USSR, following overhaul, in the late 1980s. The plane wears the standard camouflage pattern for all export MiG-23s, and IrAF insignia in their usual positions.
(Tom Cooper Collection)

Table 6: Serial Numbers

Serial numbers on IrAF MiG-23s were always applied in black on front fuselage. During the reorganization of 1988, old serial numbers were usually painted over with sand or green, and new serial numbers applied in black instead. For better understanding of serial numbering system applied on IrAF MiG-23s, the list below includes serials of two-seat MiG-23UBs as well.

IrAF Serial No.	Type	Delivery	Remarks
Pre-1988 serials			
1041	MiG-23MS	1974	1st batch
1428	MiG-23BN	1976	2nd batch
1449	MiG-23MS	1976?	Attrition replacement
1618	MiG-23BN	1977	3rd batch
1674	MiG-23UB		Captured 2003, Tammuz
1675	MiG-23UB		Captured 2003, Tammuz, c/n 1037408?
Post-1988 serials			
23001	MiG-23UB		Captured 2003, Bakr
23002	MiG-23UB		Captured 2003, Bakr
23003	MiG-23UB		Captured 2003, Bakr
23004	MiG-23UB		Captured 2003, Qadessiya
23019	MiG-23UB		Destroyed 1991, A.I.A. Talib
23020	MiG-23UB		Destroyed 1991, A.I.A. Talib
23022	MiG-23UB		Destroyed 1991, A.I.A. Talib
23023	MiG-23UB		Destroyed 1991, A.I.A. Talib
23049	MiG-23MS		Fate unknown
23070	MiG-23BN		Destroyed 1991, A.I.A. Talib
23072	MiG-23BN		Captured 2003
23081	MiG-23BN		Captured 2003
23086	MiG-23BN		Captured 2003
23103/4012	MiG-23MS		Captured 2003, Qadessiya
23104	MiG-23BN		Captured 2003
23105	MiG-23MS		Captured 2003, Qadessiya
23114	MiG-23MF		Captured 2003, Bakr
23117	MiG-23MF		Captured 2003
23121	MiG-23MF		Captured 2003
23124	MiG-23MF		Captured 2003
23126	MiG-23MF		Captured 2003, Bakr
23127	MiG-23MF		Captured 2003
23132	MiG-23MF		Captured 2003, Bakr
23134	MiG-23MF		Captured 2003, Bakr
23136	MiG-23MF		Captured 2003, Bakr
23151	MiG-23BN		Captured 2003, Bakr
23160	MiG-23BN		Destroyed 1991, A.I.A.Talib, c/n 02545
23163	MiG-23BN		Flown to Iran, 1991

IrAF Serial No.	Type	Delivery	Remarks
23166	MiG-23BN		Destroyed 1991, A.I.A. Talib
23168	MiG-23BN		Captured 2003, Bakr
23169	MiG-23BN		Flown to Iran, 1991
23170	MiG-23BN		Flown to Iran, 1991
23172	MiG-23BN		Destroyed 1991, A.I.A. Talib
23173	MiG-23BN		Sighted 1989, Fate unknown
23176	MiG-23BN		Captured 2003
23178	MiG-23BN		Destroyed 1991, A.I.A. Talib
23179	MiG-23BN		Fate unknown
23181	MiG-23BN		Destroyed 1991, A.I.A. Talib
23182	MiG-23BN		Fate unknown
23183	MiG-23BN		Flown to Iran, damaged on landing
23185	MiG-23BN		Captured 2003, Bakr
23186	MiG-23BN		Captured 2003, Bakr
23200	MiG-23ML		Captured 2003, Bakr
23255	MiG-23ML		Fate unknown
23269	MiG-23ML	1983	Impounded in Serbia, c/n 25056
23270	MiG-23ML		Captured 2003, al-Bakr
23272	MiG-23ML		Captured 2003, al-Bakr
23273	MiG-23ML		Captured 2003, al-Bakr
23281	MiG-23ML		Captured 2003, Qadessiya
23285	MiG-23ML		Flown to Iran, 1991
23286	MiG-23ML		Flown to Iran, 1991
23287	MiG-23ML		Captured 2003, al-Bakr
23294	MiG-23ML		Flown to Iran, 1991
23295	MiG-23ML		Flown to Iran, 1991
23299	MiG-23ML		Flown to Iran, 1991
23300	MiG-23UB		Captured 2003, Baghdad IAP
23306	MiG-23ML		Flown to Iran, 1991
23307	MiG-23ML		Flown to Iran, 1991

Chapter 7

Mikoyan i Gurevich MiG-25

MiG-25PD/PDS, MiG-25R & MiG-25RBT
(ASCC Code: Foxbat)

Service History

The Iraqi intentions to obtain MiG-25P interceptors were shown already in 1977, when large posters appeared in Baghdad, inviting youngsters to join the IrAF – illustrated with an aircraft that was not yet in service: a dashing MiG-25. Nevertheless, it took two years of long negotiations until the USSR agreed to deliver 12 MiG-25P interceptors and 12 MiG-25R reconnaissance aircraft to Iraq.

Foxbats began arriving in Iraq in mid-1980, packed in crates, to be put together by Soviet technicians at Tammuz AB. Priority was given to the MiG-25Ps, especially once the war with Iran began. This version entered service with the newly-established unit, No.96 Squadron, based at the same air base. Right from the start, the Iraqis were particularly proud of their "Fast One", and the unit insignia left little doubt about what kind of plane is flown by No.96 Squadron.

The unit designated to operate MiG-25Rs was also entirely new. The No.84 Squadron, though its pilots were all highly-experienced MiG-21R-fliers from the

The Iraqi MiG-25PDS (#25211) was the only IrAF MiG-25PDS captured intact in 2003. These photographs are widely available in the public. Found covered in a camouflage net, and with its wings dismantled, the plane still shows sufficient details of its colours and markings, including the serial number. (US DoD)

81

Iraqi Fighters, 1953-2003

An IrAF MiG-25PDS, as the whole fleet appeared after the reorganisation of 1988. This aircraft is painted in light grey overall, with a prominent anti-glare panel in black on the nose. Note that all dielectric panels were painted in a colour slightly darker than Dark Grey (FS25152). Insert show the insignia of No.96 Squadron as well as R-40RD and R-40TD missiles. The unit insignia has not been known to have ever been applied to any aircraft. Standard weapons configuration consisted of two R-40RDs and two R-40TDs missiles. The R-60s missiles were very seldom carried. Note that the wings of R-40 missiles usually appeared as painted in black or dark grey. They were actually not painted, but their dark blue colour came from titanium – the material from which they were made.

Overall appearance of IrAF MiG-25Rs (and later RBTs) was very similar to that of MiG-25PDS'. Major difference is that MiG-25PDS lacked the big radome for Smerch 2A radar, but it had several dielectric panels for their ELINT equipment instead. Note also a different RWR-fairing on intake, as well as a post-1988 serial number. The insert shows the position of national marking on upper side of the right wing. It was applied in the same position on the left wing, as well as undersides of both wings. The serial number of this plane is as reported to have been worn during the 1980s.

No.70 Squadron. While originally lacking attack equipment, all the Iraqi MiG-25Rs could carry the giant 5,000 litre drop tanks, called "dinghy" by their crews, because they resembled a small boat in shape and size. The 5,000-litre drop tank was carried by all IrAF MiG-25RBTs during their raids on various targets deeper inside Iran. Though No.84 Squadron became operational only after the No.96 squadron did. It became more apparent to the Iranians, flying its first operational sorties already in the spring of 1982. Initially, these concentrated on high-speed reconnaissance passes mainly over the south-western Iranian Province of Khuzestan, and Khark Island.

The IrAF was fascinated with advantages offered by MiG-25Rs. The plane could operate at altitudes over 20,000m and speeds of nearly Mach 2.8. Built-in vertical- and oblique cameras were of high quality and provided coverage of huge swaths of terrain, with a high resolution of one meter. Besides, they were equipped with SRS-16 ELINT-receivers (though no SLAR was mounted), sensitive enough to detect radar station operating inside southern Soviet Union while underway over northern Iraq! ECM-protection was provided by internal SPS-141MVG.

Though one MiG-25R was certainly shot down by Iranian F-14s over Khark Island, in 1982, in the eyes of the Iraqis this incident had little to no impact on almost daily operations. MiG-25Rs thus enabled the IrAF to establish an extensive photographic database of all major economic and military installations in western Iran.

Serial number 25105, as seen here following its capture by US troops, at al-Qadissya AB (al-Assad), in 2003. Note the unusual application of the triangle on intake sides, which is in line with aircraft, not the ground — as tradition in the IrAF. This aircraft is one of at least two IrAF MiG-25s known to have been taken to the USA. (US DoD)

Operations of MiG-25Ps were initially nowhere near as successful. As originally delivered, these aircraft were equipped with Smerch radar and old R-40R missiles only. The Iraqis persistently demanded more advanced equipment, however, and in 1983 a sizeable Soviet delegation arrived, together with a shipment of Smerch 2A low-PRF radars (working at frequencies between 9 and 9.7GhZ), TP-26 IRST-systems. Also delivered was an avionics package that brought Iraqi Foxbats up to MiG-25PDS (export standard), which was compatible with R-60MK missile, as well as slightly more advanced R-40RD and R-40TD missiles. Specifically, their export variants, working at frequencies between 9.2 and 9.4GHz. The Smerch 2A provided the pilot with search- and track modes only, and lacked the illumination mode: in attack, it continued working in pulse while guiding the missile. Its specialty was that it had a secondary, "range-only" radar, that could measure the distance to the target even in the face of heaviest jamming. This subsystem was to prove its worth beyond doubt on several occasions. All modified aircraft were designated MiG-25PDS.

The No.96 Squadron spent most of 1983 with training its pilots to use this new equipment, the following year this type was re-introduced to service. By the end of the war, Iraqi MiG-25PDS' were penetrating ever deeper inside Iranian airspace, eventually clashing several times with Iranian fighters, and claiming roughly a dozen of air-to-air kills. The first aircraft known to have been shot down by No.96 Squadron was a Grumman Gulfstream III jet, carrying the Algerian foreign minister on a secret peace mediation mission to Iran, in 1981. Other confirmed kills include at least one each against an F-4E, an RF-4E, one F-5E, and a C-130 transport. Only one MiG-25PDS was written off, when the plane damaged by a short gun burst from an IRIAF F-5E, in June 1986, made a hard landing back in Iraq.

In March 1985, in response to Iranian Scud-attacks on Baghdad, the Iraqis decided to arm their MiG-25Rs with Spanish-made bombs, and launch attacks on Tehran. A plane configured that way was presented to several Soviet experts at Tammuz AB. There followed a fierce protest from Moscow, but eventually the Soviets asked for meeting senior IrAF officers, during which they expressed

Iraqi Fighters, 1953-2003

Generally similar in colours and configuration to other Iraqi MiG-25s, #25106 was also found at al-Assad AB. These two photographs illustrate nicely the light grey overall colour of the aircraft, as well as the dark grey colour of all dielectric panels, down to the RWR-blister on intake sides, which was painted white. All the service and maintenance stencilling was in English. (US DoD)

readiness to deliver attack equipment for MiG-25Rs. Given that the Peleng-M nav/attack suite was already installed in Iraqi Foxbats, and the IrAF was already in possession of a stock of some 1,000 FAB-500T bombs, all the Soviets had to do was deliver multiple-ejector racks. Within few days, the first few Iraqi MiG-25RBTs appeared, capable of carrying a maximum load of eight bombs: two under each wing and four under fuselage. On 13 March 1985, the first raid against Tehran was flown by nobody else but CO No.84 Squadron himself.

In the following weeks, such operations were intensified, as the number of available MiG-25RBT converts increased. In order to better cover various targets in Iran, they were deployed at several airfields. A contingent of two Foxbat-Bs was deployed to Hurrya AB, from where they raided Tabriz, Qazvin, Karaj, Rasht and Hamedan, usually carrying four FAB-500Ts only. A third contingent of two aircraft was deployed to Abu Ubaida AB, and these operated against Esfahan, Shiraz, Bushehr and Khark, depending on the range to the target. Later in the war, most of the MiG-25RBT operations were shifted to Ali Ibn Abu Talib AB. The HQ of No. 84 Squadron remained at Tammuz AB. The main thrust was aimed at Tehran, and the Qom, both of which were subjected to raids by day and night. Tehran was usually attacked by single MiG-25RBTs carrying four FAB-500Ts; for raids against Qom, up to eight FAB-500Ts were carried.

After crossing the border, Iraqi Foxbats got usually underway at a speed of Mach 1.2 trough 1.8, and level 20,000m, accelerating to Mach 2.5 or more for their attack runs. Bombs were released from a horizontal distance of 47km to the target. The precision was not particularly high, but Iraqis understood the problem and accepted the fact that the Peleng-M lacked pin-point accuracy, having a calculated error point (CEP) of 500m. This was considered a good figure since the original Soviet MiG-25RBs were supposed to deliver nuclear weapons only.

By the summer of 1986 the IrAF concluded, it could not sustain the sortie rate with the number of MiG-25RBTs available. Additional aircraft were requested from Moscow. This time, the Soviets reacted in a very strange way, flying a batch of ten additional MiG-25PDs and six MiG-25RBTs to Iraq via Turkey. All communications were in Russian: obviously, they wanted to deliver a strong political

Two from a famous series of photographs showing the unearthing of MiG-25RBT #25107, near al-Qadissya AB, in June 2003. Surviving in quite good condition, and having most of its avionics intact. This plane was later taken to the USA. (US DoD)

message to Ankara, Washington, and Tehran. This delivery increased the number of available MiG-25PDS to 20, and MiG-25RBTs to 17.

There were problems with pilots as well, the strain on them was enormous. They had a special diet, and were later to grow older rather quickly, with wrinkles and shabby yellow faces. The rate of cardiac arrests was higher than in other fighter-pilot communities, reportedly due to sustained Mach 3 flying. Such problems eventually necessitated increasing the rate of pilots per aircraft in each unit from 1.5:1, to 2:1 during the war. The requirement to keep the level of proficiency high necessitated also a relatively large number of two-seat aircraft: at least six MiG-25Pus were purchased over the time.

Overall, the IrAF should have lost just one more MiG-25RBT during the war. In addition to the example shot down by F-14s in 1982, another Foxbat was splashed by Iranian SAMs in February 1987. Iranian sources claim for up to 12 Iraqi MiG-25s, several of which are well-documented, plus two Soviet-flown MiG-25BMs. Some of Iranian claims are supported by US intelligence reports. Due to the lack of corresponding Soviet reports, it is practically impossible to establish with 100 percent certainty whether the Soviets have ever operated any MiG-25s out of Iraq, or on behalf of IrAF. It is possible that some kind of secret operations were undertaken. Surely, all the aircraft of No.84 and No.96 Squadrons were always manned by Iraqi personnel. There were unconfirmed UN reports that indicated at least one foreign pilot being granted permission to fly one of MiG-25RBs.

By January 1991, the IrAF still had 19 MiG-25PDS/PD interceptors, as well as 16 MiG-25R/RBTs, based at Tammuz AB, in their famous hexagonal shelters of immense size. During the War against the Gulf Coalition, the No.96 Squadron did what was possible to remain operational under most difficult circumstances, and most of the effects of its carefully planned missions remain unrecognized until today. Hampered by repeated strikes against Tammuz AB, pilots of this unit had to abort a number of sorties while rolling out, their aircraft being damaged by mines strewn from JP.233 dispensers carried by RAF Tornados (one MIG-25PDS was hit by these in the first night of the war, and its pilot badly injured), or prevented from reaching runways. The few Foxbats that did manage to take off and received a useful intercept vector from ground stations have caused considerable problems to several US formations. On the first night of the war, a MiG-25PDS flown by a

Iraqi Fighters, 1953-2003

MiG-25RBT #25109 differed from the other Iraqi recce-Foxbats found in 2003, in having no RWR-blister but slightly larger national insignia on intake sides, and a slightly deeper anti-glare panel on forward fuselage. Contrary to other Foxbats captured by US troops, it has also still had its wings and even multiple-ejector-racks for FAB-500T bombs attached. (US DoD)

These two photographs show the wreckage of one of four MiG-25RBTs found burned on junk yard near Tammuz AB, specifically the underside of the left wing, together with a multiple ejector rack. Of interest is the national insignia, pointing outwards, as usually on IrAF Foxbats. (Tom Cooper Collection)

Two wrecked MiG-25PDS', found by US troops near al-Qadessiya AB, in 2003. Of special interest is the large chaff & flare dispenser carried on upper surface of the aircraft seen to the right. Unrecognized even by US intelligence before 1991, these dispensers could carry a large number of flares, which proved to be very effective in decoying even such advanced air-to-air missiles such as the AIM-7M Sparrows and AIM-9M Sidewinders missiles. The wing of the same wreck also shows traces of national insignia. (DoD)

young captain approached unrecognized (but, probably, not "undetected") a USN strike package from USS Saratoga, and shot down an F/A-18C Hornet despite intensive jamming and approach at Mach 1+ flight at medium level. On the third day of the war, when the IrAF launched its only attempt to seriously challenge the Allied control of its skies, No.96 Squadron clashed with USAF F-15Cs, losing two aircraft in the process. Based on this experience, another carefully planned mission was launched on 30 January 1991, after which the IrAF claimed one USAF F-15C as hit by R-40R missiles. Neither of these two kills was initially considered as "confirmed" by IrAF. The Iraqis did not know about the loss of the USN Hornet until 1993, and have declared that kill for confirmed only in 1995. Something similar happened with the claim for the Eagle, and this kill was also confirmed only years later, in part through the work of the Iraqi intelligence, and in part through official contacts between IrAF officers and US military intelligence officers working as UN inspectors in Iraq.

When IrAF resumed flying operations, in 1992, the No.96 Squadron remained at Tammuz AB, but No.84 Squadron was moved to al-Qaddessya AB (better known as al-Assad AB).

Two MiG-25PDS' were shot down in air combats and a total of 18 various Foxbats destroyed on the ground during the war, thus leaving the IrAF with only 15 airframes of all three versions. Nevertheless, the No.84 and No.96 Squadrons were some of the most active IrAF units during the confrontation with US and British air forces over northern and southern Iraq, in the 1990s. MiG-25PDS were challenging the so-called no-fly zones already in the late 1992, when one was claimed as shot down by USAF F-16s – though no such loss is known in Iraq. MiG-25RBTs often flew photo-reconnaissance sorties inside Saudi and Kuwait airspace.

From 1998 onwards, both units became involved in a number of experiments, when the IrAF launched a project of re-vitalisation of its interceptor fleet. One of the ideas was to mate a MiG-25RBT with French-made Baz-AR anti-radar missiles (for details see Chapter 9 Mirages), enabling the plane to attack one of US or British AWACS aircraft. Though this project was abandoned at an early stage, one MiG-25RBT was stripped down of all the avionics in preparation for this modification – only to be returned to service in its original configuration.

One of the final missions flown by Iraqi MiG-25s occurred on 23 December 2002, when a MiG-25PDS shot down an USAF RQ-1 Predator armed UAV, south of Baghdad.

Camouflage, Markings & Serial Numbers

All MiG-25s delivered to Iraq have been got a coat of heat-resistant light grey colour overall, with prominent anti-glare panels in black, in front of the cockpit. Underside and sides of fuselage in front of engine exhausts was painted in heat-resistant aluminium colour. Fin flashes were worn on all Foxbats, and triangles applied in six positions. Those on the upper and lower wing surfaces were canted some 30 degrees outwards. Those on fuselage sides were applied on intakes, instead of rear fuselage. Drop tanks were always left in bare metal overall. Serial numbers were applied in black below the cockpit, the construction number was sometimes applied on the top of the fin, in the same fashion like on Algerian MiG-25s. No aircraft is known to have got any special markings. All maintenance stencilling was in English language.

Table 7: Serial Numbers

Serial numbers were applied in black on the front fuselage. Given that the number of MiG-25s delivered to Iraq remains a matter of controversy, and nothing is left of corresponding IrAF records, the following list mentions also aircraft with unknown serial numbers, known to have been lost at various points in time, so to illustrate the approximate number of MiG-25-airframes delivered to Iraq.

IrAF Serial No.	Type	Delivery	Remarks
25001	MiG-25PU	1981	Captured 2003, Qadessiya
25002	MiG-25PU	1981	Captured 2003, Qadessiya
25003	MiG-25PU	1981	Reported derelict at TFB.3, Hamedan, Iran
2500?	MiG-25PU	1981	Crashed over western Iran, 1991
2500?	MiG-25PU	1981	Crashed due to fuel starvation, 2000
2510?	MiG-25R	1981	Shot down 1982, over Khark
252??	MiG-25RBT	1981	Shot down 1987, central Iran
2125	MiG-25RBT	1981	Pre-1988 serial
25103	MiG-25RBT	1981	Wreckage at Tammuz, 2003
251??	MiG-25RBT	1981	Wreckage at Tammuz, 2003
251??	MiG-25RBT	1981	Wreckage at Tammuz, 2003
251??	MiG-25RBT	1981	Wreckage at Tammuz, 2003
251??	MiG-25RBT	1981	Wreckage at Tammuz, 2003
25104	MiG-25RBT	1981	Destroyed 2003, Tammuz
25105	MiG-25RBT	1981	Captured 2003, taken to USA
25106	MiG-25RBT	1981	Captured 2003, Qadessiya, fate unknown
25107	MiG-25RBT	1982	Captured 2003, taken to USA
25108	MiG-25RBT	1982	Sighted 1989, fate unknown
25109	MiG-25RBT	1982	Captured 2003, Qadessiya, fate unknown
252??	MiG-25PDS		Written off 1986, after interception by F-5E
252??	MiG-25PDS		Hit JP.233 mine, Tammuz AB, 1991
252??	MiG-25PDS		Shot down 1991, western Iraq
252??	MiG-25PDS		Shot down 1991, western Iraq
252??	MiG-25PDS		Shot down 1992, southern Iraq
252??	MiG-25PDS		Destroyed 2003, Qadessiya
252??	MiG-25PDS		Destroyed 2003, Qadessiya
252??	MiG-25PDS		Destroyed 2003, Tammuz
25201	MiG-25PDS	1980	Destroyed 2003, Qadessiya
25202	MiG-25PDS	1980	Destroyed 1991, A.I.A. Talib
25204	MiG-25PDS	1980	Destroyed 2003, Tammuz
25205	MiG-25PDS	1980	Destroyed 2003, Qadessiya
25211	MiG-25PDS		Captured 2003, Bakr, since destroyed
25212	MiG-25PDS		Captured 2003, fate unknown
25216	MiG-25PDS		Last seen operational in 1991

Chapter 8

Mikoyan i Gurevich MiG-29

MiG-29
(ASCC Code: Fulcrum)

Service History

The IrAF was among some of the first countries outside the USSR to learn about the existence of MiG-29. They became the first export customer for the MiG-29, issuing an order for 32 single-seat and four two-seater models in late 1985. Early in 1986, a group of pilots and technicians were dispatched to Soviet Union for training, but the first MiG-29s were delivered in late 1987. They entered service with the re-established No.6 Squadron, based at Tammuz AB. This unit began flying operational sorties on 17 April 1988, and from then on participated in all the major Iraqi operations during the last four months of war. Though there were few opportunities when MiG-29s were scrambled, they fought no air-to-air battles with any Iranian fighters.

This MiG-29 serial number 29062 was seen during defence exhibition in Baghdad, in 1989. Despite its serial number, the camouflage pattern reveals it also is from the first batch of 16 MiG-29s delivered to IrAF. Of interest is the unusual distribution of national markings, not only fin flashes. Also triangles were applied on fin. There was no national insignia on upper wing surfaces.
(Christopher Foss)

Iraqi Fighters, 1953-2003

One from the first batch of 16 MiG-29s delivered to IrAF is serial number 29062. The plane served with No.6 Squadron, the insignia of which is illustrated in the left upper corner, together with insert showing details of the serial applied on intake. All maintenance instruction stencilling was in Russian.

MiG-29 "9.12B" #29040 also belonged to the first batch as well, and has had its serial number applied with help of air brush. Note the lack of national markings on upper wing surfaces.

Another photograph of the same aircraft at the same event, illustrating position of national markings to advantage. Note the traditional inclination of fin flash, even if this was very subtle when compared with MiG-23s or some other types. (Christopher Foss)

Mikoyan i Gurevich MiG-29

Like most of early export-MiG-29s, Iraqi Fulcrums were equipped with the RLPK-29 weapons system, including the X-band N019 Rubin look-down/shot down radar, OEPrNK-29 optical set with OEPS-29 KOLS IR-tracker and laser rangefinder, as well as the NSC-29 helmet mounted sight (HMS). The N091 radar could track up to ten targets simultaneously (when working in the "Head On" mode), out to a range of 70km in search, and 60km in the track mode. The KOLS system had a maximum search range of 15km and tracking capability out to 12km, an integrated laser rangefinder had a range of 6.5km in optimum conditions with a forward field of view of 60 degrees in azimuth, and 30 degrees in elevation, and a capability to scan the entire envelope within only 3.5 seconds.

Main armaments of IrAF MiG-29s were R-27R semi-active radar-homing-, and R-60MK IR-homing missiles. Though in theory the type could carry bombs and unguided rockets, the pilots spent very little time training for air-to-ground tasks.

Original Iraqi intention was apparently to equip two units with the type, but such plans were never realized. In fact, the Znamya Truda factory put together a total of 70 MiG-29s for Iraq, but by the time the Iraqis learned about the existence of an even more advanced and superior type: the Sukhoi Su-27 was promising to become much more suitable to IrAF requirements and an acquisition of significant numbers was planned. Besides, the Iraqis were deeply involved in development of Mirage 2000D and associated weapons (including a conventional-version of ASMP supersonic cruise missile, and MICA air-to-air missile), and thus the IrAF was slow to collect all the MiG-29 it had ordered. In fact, deliveries were stopped already by the mid-1989, and No.6 Squadron remained the only unit equipped with the type.

As of January 1991, IrAF had 33 operational MiG-29 single-seaters, mainly based at Tammuz AB. Only two aircraft were forward deployed to Jaliba airfield, in southern Iraq. The type was heavily involved in air-to-air combats with Coalition aircraft within the first few days of the war, and USAF pilots claimed up to

MiG-29 #29040, as seen in early 2006, near former Tammuz AB. Most of blue grey colour on upper surfaces was washed away by sun, sand and rain, but camouflage pattern is recognizable enough to show the plane as belonging to the second batch supplied to Iraq. Of interest is the position of national insignia on lower surface of starboard wing, inclined some 30 degrees outwards.
(Tom Cooper Collection)

91

Iraqi Fighters, 1953-2003

Close up photograph of serial number #29040, showing that it was applied with brush, probably with some help of adhesive tape. (Tom Cooper Collection)

The serial number on #29062 was obviously applied with air brush. (Tom Cooper Collection)

Mikoyan i Gurevich MiG-29

eight IrAF MiG-29s as shot down. In fact, only three were lost in air combats. In turn, they claimed damaging a B-52G and an F-111F, as well as a shooting down an RAF Tornado, on 19 January 1991. Seven other MiG-29s were destroyed on the ground, and four flown to Iran. Two additional MiG-29s were lost in accidents between 1987 and 2003.

Following the war, the IrAF MiG-29 fleet was in very poor condition, and by the mid-1990s it was practically grounded, mainly due to engine-related problems. Another plane has been lost when its pilot attempted to fly to Saudi Arabia, in late summer of 1995, and thus only 13 single- and three two-seaters remained. The number of serviceable airframes decreased to eight as of February 1996, and these were subsequently stored. In the early 2003, the IrAF began receiving some smuggled spared parts for engines, but never introduced them to service.

A view of upper surface of port wing reveals that no national insignia was ever applied there. Most of the blue-grey colour was washed out, but its traces can still be recognized.
(Tom Cooper Collection)

Camouflage, Markings & Serial Numbers

Iraqi MiG-29s were painted in Light Grey (FS26373) overall, with wide splotches of Grey-Green (FS35352) on upper surfaces, around the cockpit, wing-tips and fins. The later colour tended to swiftly deteriorate into various shades of blue-grey under effects of sun and elements. Small difference in camouflage pattern between the first and the second batch of single-seaters (16 aircraft each) can be seen on various photographs. The first batch has had two almost vertical splotches of blue-grey on outside surfaces of fins. The second only one, usually covering the front part of the fin. National markings were applied on fins and lower wing surfaces only. Serial numbers in black on intakes. No special insignia is known to have been worn.

National marking on lower surface of #29062's starboard wing. It was pointing outwards at something like 30°. Note that the outside border of black outline was quite uneven, even if applied with help of adhesive tape.
(Tom Cooper Collection)

Probably strangest of all known cases was the application of national marking on lower surfaces of #29040's starboard wing. The marking was painted very far inwards, right besides the launch rail for R-27 missiles, and pointing straight ahead.
(Tom Cooper Collection)

National marking on lower surface of #29040's port wing was applied in a particularly awkward manner, pointing inwards! Note the different shade of green colour, as well as a finer black outline than on #29062.
(Tom Cooper Collection)

Iraqi Fighters, 1953-2003

In their intended role, IrAF MiG-29s were pure interceptors, planned to operate in conjunction with ground-based radar stations and Adnan-2 AEW-aircraft, of indigenous Iraqi construction, based on Il-76 airframe. Though two Adnan-2s had been built by 1991, none were operational at the times of war with Gulf Coalition, and both were evacuated to Iran. The two MiG-29s seen on this photograph belonged to the second batch supplied to Iraq, and otherwise wear standard national markings.
(Tom Cooper Collection)

Table 8: Serial Numbers

Serial numbers were applied in black on intake sides only, sometimes with brush, in other cases with help of air gun and stencils. There were considerable gaps between the serial numbers, and then the sequence of known ones goes well beyond the number of aircraft known to have been delivered. The serial numbers of MiG-29s from second batch were selected within the same range, filling various gaps.

IrAF Serial No.	Type	Delivery	Remarks
29004	MiG-29UB (9.51B)	1987	Flown to Iran, 1991
29030	MiG-29 (9.12B)	1987	Captured 2003
29032	MiG-29 (9.12B)	1987	Flown to Iran, 1991
29038	MiG-29 (9.12B)	1987	Flown to Iran, 1991
29040	MiG-29 (9.12B)	1987	Captured 2003, c/n 21830
29044	MiG-29 (9.12B)	1987	Flown to Iran, 1991
290??	MiG-29 (9.12B)	1988	Wreck in 2003, c/n 21843
29050	MiG-29 (9.12B)	1988	Captured 2003
29056	MiG-29 (9.12B)	1988	Captured 2003
290??	MiG-29 (9.12B)	1988	Wreck in 2003, c/n 22311
290??	MiG-29 (9.12B)	1988	Wreck in 2003, c/n 22312
29059	MiG-29 (9.12B)	1988	Captured 2003
29060	MiG-29 (9.12B)	1988	Last seen 1989
29062	MiG-29 (9.12B)	1988	Captured 2003, Tammuz, c/n 22994
29063	MiG-29 (9.12B)	1988	Captured 2003, Tammuz
29072	MiG-29 (9.12B)	1988	Captured 2003, Tammuz

Chapter 9

Dassault Mirage F.1

Mirage F.1EQ, F.1EQ-2, F.1EQ-4, F.1EQ-5 & F.1EQ-6 & F.1BQ

Service History

The history of Dassault Mirage F.1 in service with IrAF is perhaps the best-known, and the most prominent one (at least outside Iraq). Several different versions of F.1EQ entered service in the Iraqi AF, and although it was far outnumbered by types supplied by the USSR, this type became the most important strike asset. They proved definitely to be one of the most effective interceptor types in Iraqi hands. In service with IrAF, the F.1 became both, legendary and famous - foremost due to the Iraqis treating the type as flown by the "chosen ones", by carefully selected pilots, and maintained by best engineers and technicians.

At least as important is the fact that combat experiences from Iraqi Mirage F.1EQs were of immense importance for the subsequent development of Mirage 2000, and a full range of French-made avionics- and weapons systems.

Nicely illustrating the Kaki colour with its fresh green touch, F.1EQ #4010 is seen during pre-delivery testing, in France, in 1980. Of interest is the UHF-filet on the joint of fin's leading edge and fuselage. Originally delivered as a "multi-role" fighter, Mirage F.1EQ served its first years with IrAF almost exclusively as interceptor, and was involved in dozens of air combat missions with Iranian interceptors. Later on, it was heavily deployed for air-to-ground missions. (Dassault Aviation via Tom Cooper).

Iraqi Fighters, 1953-2003

Probably the most famous IrAF Mirage F.1EQ was #4014, well known in the West from a number of photographs showing it while being used for weapons compatibility trials of various weapon systems configurations. Following its delivery to Iraq the F.1EQ became a kind of a personal mount of XO No.79 Squadron, Major M. during the period September 1981 to June 1982, along with several other pilots claimed a total of 14 kills against Iranian interceptors. This particular plane made most of the kills with Super 530F-1 missiles, the rest were with Magic missiles. This while operating out of Wahda AB, near Basrah. Subsequent official inquiries of HUD video and radar bands revealed that five of the claimed kills could be considered as "confirmed". Nevertheless, Major M has already applied 14 kills on the Mirage, and that is how it remained until captured by US troops at Qadessiya AB, in 2003. In the late 1990s, this plane was one of few surviving F.1EQs that was upgraded with Cyrano-IVM radar. For this achievement, its ground crew was granted official permission to apply the insignia of "Saddam Fedayeen" on the left side of forward fuselage, below the cockpit.

From December 1982, Mirages of No.79 Squadron began deploying AS.37D Bazar anti-radar missiles as well. According to recollections of former IrAF pilots, each aircraft carried only one Bazar under centreline pylon, which was quite a tight fit, leaving just a few centimetres between fin-tips and surface when aircraft was on the ground.

Dassault Mirage F.1

One of five Super Etendards leased to Iraq from October 1983 was #65, shown here together with a group of IrAF and French Air Force pilots. The aircraft wears the then standard camouflage pattern of gris marin foncé (dark sea grey) over, and Insignia White (FS17875) underneath. Relationship between the two air forces was very close at the time, and several French Air Force pilots worked with the IrAF, one of them spending so much time in Iraq that he earned himself the nick-name "Arab".
(Ahmad Sadik Collection)

Iraq was looking for a ground attack aircraft with most advanced avionics and weapons since 1969. A search strongly influenced by the 1967 War, when French-made Mirages flown by the Israelis appeared as a kind of a "key to victory" to many in the Middle East. The reputation of Mirage was immense. During a meeting between the Egyptian and Israeli Ministers of Defence, following the signing of a peace accord between Egypt and Israel, in 1977, the Israeli minister asked his counterpart about the latter's opinion on the best thing the Israeli Armed Forces have ever had. The answer was swift and firm, the Mirage. Without any surprise by the end of 1967, there were first rumours about negotiations between Baghdad and Paris. These ended without substantial results. It was only the Iranian-supported Kurdish insurgency in northern Iraq, from March 1974 until March 1975, that resulted in the Iraqi decision to negotiate seriously with the French.

During this crisis, Moscow exercised immense pressure on Baghdad, forcing the Iraqi Government to end the war through negotiations, even under unfavourable terms, if required. The bargain chip in Soviet hands were arms and spares, and Iraqis realised they were in a need of a different source, in order to avoid a similar situation in the future. Thus began the "March on Paris".

In 1976, a high-level Iraqi political and military delegation visited France. They paid a visit to the French Air Force test centre, where a flight demonstration of Mirage fighters, as well as ground inspection were staged. Following short negotiations, a principal agreement was reached for procurement of Mirage F.1s, as well as various air defence assets. Final negotiations resulted in an order for 36 aircraft, and 36 options, together with associated air-to-air missiles. This was signed in June 1977, and designated "Project Baz" ("Baz" means "Falcon" in Arabic).

97

Iraqi Fighters, 1953-2003

As delivered to Iraq, all Mirage F.1EQs and F.1BQs wore this disruptive camouflage pattern of sand and "kaki". The later colour was rather light-sensitive and prone to look differently under different light. This prompted many observers to believe that Iraqi Mirages were actually camouflaged in sand and dark earth instead sand and kaki. The F.1 No. 4009 is depicted carrying a single RP.35 drop tank under the centerline pylon and a BG-66 Belouga CBU on a TER under the inboard underwing pylon. This was one of often seen configurations once the type was deployed as fighter-bomber from 1984 onwards. The wing-tip mounted Matra R.550 Magic missiles were always carried, but are "removed" from this drawing in order to show the details of inboard installations and camouflage.

In 1984, the IrAF learned from Morocco about the US-made Mk.84 bombs equipped with the so-called "Daisy Cutter" fuses detonating approximately one meter above the ground. While the Moroccans were not particularly pleased by the weapon's performance in West Sahara, the Iraqis purchased Mk.84 bombs from Saudia Arabia and armed them with South African "Jupiter" fuses. Such weapons were deployed with good effect against large formations of Iranian infantry, starting from 1984.

Period I: Battlefield Air Superiority

The first version to enter service with IrAF was designated F.1EQ: E stood for multi-role (to distinguish this version from interceptor F.1C), and Q for Iraq. The first 16 F.1EQs were actually little more but F.1Cs, with addition of HF extra dorsal filet aerial at the forward joint of fin and fuselage, and BF RWR. Though mainly intended to serve as interceptors, equipped with Cyrano IV-1 radar and Super 530F-1 air-to-air missiles, various weapons and weapons configurations were tested on early Iraqi Mirage F.1s.

The Cyrano IV-1 radar was an X-band, monopulse radar, with coaxial magnetron as the signal generator. It could work in low- and medium PRF-modes, on frequencies between 9,000 and 9,600MHz, and could provide range displays for 110km/68nm, 65km/40nm, and 28km/17nm, with scan volume of 60 degrees to each side in azimuth, and +/-30 degrees in elevation. Not as optimised as contemporary US-made systems, Cyrano IV-1 was highly automated. Once an airborne target was detected, the pilot only needed to place the cursor on it. The radar would automatically change into tracking mode, and remained locked-on until target manoeuvred outside the envelope, was shot down, or the pilot stopped interception. The weapons system was providing the pilot with the earliest, optimum, and latest time-point for firing the selected weapon. Another important feature was an air-to-ground mode, called "designation". This was used to measure the exact distance between the aircraft and ground target. Computer of weapons system would then calculate the optimum release point of selected weapon, which resulted in particularly good accuracy when delivering general purpose bombs. Mirage F.1EQs have also got a radar altimeter, and wiring for carriage of COR-2 reconnaissance pod (even if none of these were delivered before mid-1984). Later on Harold and Raphael TH reconnaissance pods were supplied as well.

To complete the original order, already in 1978, the IrAF issued an order for a follow-up batch of 16 F.1EQ-2s, equipped with slightly improved Cyrano IV-2 radars, upgraded on Iraqi request with ground mapping and low-altitude modes. A year later, Iraq exercised the option for additional airframes. The order for 28 Mirage F.1EQ-4-200s was made to be equipped with in-flight refuelling probes, Cyrano IV-3 radar (including all the improvements from IV-1 and IV-2) and SOCRAT 6200 VOR/ILS navigation system, as well as their structure stressed for the carriage of the huge RP.35 "Irakien" drop tank (capacity of 2,200 litres), and Douglas/Intertechnique in-flight refuelling store.

Through the program, Mirages were test-flown and officially handed over to IrAF well in advance before actually being flown to Iraq. The first single-seat F.1EQ, for example, made its maiden flight on 28 May 1979, and was handed over to IrAF in April 1980, at Mont-de-Marsan, where also all the Iraqi pilots – mainly experienced former MiG-21-fliers – for the type were trained.

The first four Mirage F.1EQs reached Iraq almost one year later, between 28 and 30 January 1981, via Italy, Greece, Cyprus, and Turkey. Next six followed two days later. The fleet was right from the start based at the newly-built Saddam Air Base (better known as "Qayyarah West"), which became the main hub for all IrAF Mirage-operations.

Iraqi Fighters, 1953-2003

The only Mirage F.1EQ-4 known to have worn kill-markings was #4506, though details about corresponding air-to-air combats remain unknown. The plane is shown with three bomb-types used frequently during the war with Iran, including SAMP 400 BL70 general-purpose bombs mounted on "surfboard" under the centerline pylon, SAMP 400 Type-21C (left lower corner); and SCB CBU, a South-African-made CB470 cluster bomb, designed on drawings of an IrAF officer, Salah. This aircraft was captured in 2003, and subsequently taken to the USA for evaluation.

A Mirage F.1EQ-4 of No.81 Squadron depicted in the heaviest configuration ever carried during the war with Iran. Here carrying a giant 2,000-litre "Irakien" drop tank under the centerline pylon, developed in response to special Iraqi requirement. Two RP.35 drop tanks were carried under inboard underwing pylons, and Magic missiles on wingtips. Two SAMP 250 Type 21C bombs could still be added to outboard under-wing pylons. Further a selection of equipment & weapons is shown as deployed by Iraqi Mirages (from left to right): SICOMOR chaff & flare dispenser, Remora ECM-pod, Caiman offensive stand-off jammer, and Douglas refuelling pod.

The first IrAF Mirage F.1EQ-unit became the No.79 Squadron, established in May 1981 under Lt.Col. R., and tasked with medium altitude air defence and photo reconnaissance. Later on, low altitude interception and electronic warfare was added to its duties. As soon as the unit was declared operational, detachments of Mirage F.1EQs began operating from various airfields mostly in southern Iraq, such as Wahda AB, from where they usually launched interceptions of Iranian fighters mainly underway over the southwestern Iranian Province of Khuzestan, where heaviest air battles occurred. Net result of this modus operandi was a whole series of fierce air-to-air battles, mainly fought at medium ranges, that lasted from September 1981 until the mid-1982, and involving Mirage F.1EQs from No.79 Squadron. While the final results of many of these engagements are still a matter of dispute, Iraqi Mirage-pilots claimed some 25 kills in this period, losing several of their own in return.

In the end by 1983 the Iranians gave up contesting the IrAF in the air over the southern battlefields, and instead resorted to ground-based air defence measures. Meanwhile, as additional aircraft became available, a second unit was established, No.91 Squadron. This was mainly tasked with air defence of Mosul area, relieving MiG-21s of No.9 Squadron, which was re-located to southern Iraq. No.91 Squadron was also the first unit to start training for air-to-ground missions.

Period II: Multi-Role

Built-in armament of Mirage F.1EQs consisted of two DEFA 553, single-barrel, gas-operated revolver cannons, with firing rate of 50rds/minute, and capacity of 135 rounds per gun. Matra R.550 Magic Mk.Is were delivered to Iraq already in 1980, and by 1983, also a total of 352 Matra Super 530F-1s arrived as well. Quite a number of the later were spent in mentioned air battles, in late 1981 and through 1982, but the usually reported figure of 100 rounds fired for 35 scored kills is wildly exaggerated, as Iraqi Mirage pilots did not claim as many kills during the whole war.

Principal air-to-ground armament included Matra F4 rocket pods for 18 SNEB or Multi-Dart 68 rockets calibre 68mm, used very rarely. Matra BLG-66 Belouga cluster-bomb units (490kg), as well as SAMP Type 21-C 250 and 400kg general purpose bombs were used in large numbers from 1984 onwards, and Belouga proved an excellent anti-personnel weapon.

Anti-radar missiles were deployed more often during the first few years of Mirage's service in Iraq. Already in 1979, the Iraqis tested the French-made AJ.168 anti-radar missile in France. After a number of trial firings, two batches of 30 missiles each were ordered, the first being delivered already in the late 1979, and the other in mid-1981. Officially designated AS.37D – but called BAZAR – a designation based on Iraqi name for Mirage F.1 and AR for "anti-radar". These weapons were deployed in combat for the first time on 26 December 1982, when three Iranian radar stations in Khuzestan were attacked. In January 1983, No.79 Squadron flew its first electronic-reconnaissance (ELINT) missions, Mirages carrying the Syrel ESM-pods for this purpose, and from July of the same year also the first close-air-support missions, during which Belouga CBUs were deployed in considerable numbers. These operations signalled a significant change in the way the IrAF Mirage fleet has been tasked.

Iraqi Fighters, 1953-2003

No. 4528 shown as Iraqi Mirages usually looked like during test-flying in France and transferred to Iraq. Although serial numbers and identification codes were already in place, no Iraqi national markings were applied. Shown below the aircraft are a Matra F4 rocket launcher (left), and a version that combines the launcher with practice bomb dispenser.

Being used as interceptors and fighter-bombers, the Iraqi Mirages were also deployed frequently for reconnaissance missions, equipped with COR-2 recce pods (left) as well as Raphael side-borne-looking radar (SLAR), as shown mounted to the centre-line pylon of #4562 depicted here.

Depicted carrying an ATLIS designator pod as well as a SAMP 400 LGB, an adorned with two kill markings, the Mirage F.1EQ-5 #4570 was a personal favourite aircraft of the CO No.81 Squadron. Mirages from this unit were the only ones "wired" for ATLIS pods, LGBs and AS.30Ls. They saw extensive service in anti-ship role. They were used on long-range raids against all sorts of targets around western and southwestern Iran, including the raid against construction site of the Iranian nuclear reactor in Bushehr.

Dassault Mirage F.1

Mirage F.1EQ #4028 here seen carrying a Syrel ELINT-pod under the centerline pylon, two RP.35 drop tanks and, wing-tip mounted Magic missiles. Main feature of this pod, introduced to service in 1982 in cooperation with the then leading French specialists in this field, was its capability to transmit data in real-time, providing exceptionally precise Iranian electronic order of battle. This in turn helped IrAF plan most of its long-range raids deep into Iranian airspace.
(Ahmad Sadik collection)

Mirage F.1EQ #4008, seen carrying a COR-2 reconnaissance pod. Equipped with the later, the IrAF Mirage F.1EQ-fleet became a powerful reconnaissance platform during the long war with Iran, in the period 1981-1983. Later on, MiG-25Rs took over this role, and the carriage of COR-2s occurred very rarely.
(Dassault Aviation via Tom Cooper)

An IrAF Mirage F.1EQ – apparently #4021 – seen on take-off, armed with Super 530F-1 and R.550 Magic Missiles. Of interest is the drop tank - left in "bare metal" overall. Many of drop tanks delivered with subsequent F.1EQ-4 version were painted in "sable" (sand) overall.
(Ahmad Sadik Collection)

Iraqi Fighters, 1953-2003

Starting with Mirage F.1EQ-5 #4571, some of Iraqi Mirages were camouflaged in dark grey over and Insignia white under for anti-shipping operations. The aircraft depicted carrying — what was then known as an "Iraqi Exocet", the second version of AM.39. A special pylon mounted under the centerline of the fuselage was required for this purpose. When undertaking anti-ship operations over the Gulf these planes usually carried RP.35 drop tanks and ECM-pods (foremost Remoras) under wings. Such a load slowed the cruising speed down to barely 350 knots, and exposing the aircraft to Iranian interceptors for longer periods of time.

The five Super Etendards leased to Iraq from 1983 until 1985 wore the then standard Aéronavale's (French Naval Aviation) camouflage pattern, consisting of gris marin foncé (dark sea grey) over, and Insignia White (FS17875) below. Included were Super Etendards -numbers 64, 65, 66, 67, and 68. Their IrAF serial numbers were 7445, 7446, 7447, 7448, and 7449. Number 7447 being considered as the most remarkable by its crews. It remains unknown whether this is due to it logging the most sorties or the highest number of Exocet launches. The few photographs showing them in full insignia do not reveal if any national markings were applied on upper wing surfaces. Given the application of large triangles high on rear fuselage this is rather unlikely. Fin flashes (with strong inclination) were applied on fins. (Artwork by Ugo Crisponi)

Dassault Mirage F.1

No. 4510 was a Mirage F.1EQ-4 — the first version optimised for air-to-ground operations, but frequently used in the an aerial tanker role. It carried the Douglas "buddy-buddy" refueling tanks. All maintenance stencils on Iraqi Mirages were in English. The serial number was already applied, but national markings were not, pending delivery to Iraq.
(Dassault, via Michel Benichou)

In-flight refuelling capability was one of major reasons behind the orders for F.1EQ-4s. It proved to be a "force-multiplayer" especially during the later stages of the Iran-Iraq War. This enabled the Iraqi Mirages to reach targets considered "far beyond the reach" of IrAF previously. This photography presents a view from the rear cockpit of an F.1BQ during a training flight in France, and shows the IFR-probe already inside the basked streamed out of a Douglas pod, carried by one of Iraqi F.1EQ-4s. (Dassault, via Michel Benichou)

Iraqi Fighters, 1953-2003

In 1988, Mirage F.1EQ-5 #4563 was modified to test a combination of ATLIS laser designator with Kh-29L air-to-ground laser-guided missile. Plenty of mechanical, electronic and electrical modifications were required to make Mirage compatible with the Soviet-made weapon, and this "prototype" remained the only F.1EQ to ever be modified that way.

Mirage F.1EQ-6 #4655 as it went to a war with Iran in the early 1988, armed with Super 530D and Magic Mk.1 missiles. Delivery of a trial batch of Super 530D missiles to Iraq remains largely unknown in public, but was confirmed by former IrAF pilots and Iranian intelligence sources. Remnants of transfer code "Y-IBLU", used as radio-call sign during delivery flight from France to Iraq, can be seen in addition to usual set of markings. This Mirage survived the war but was subsequently flown to Iran.

Dassault Mirage F.1

Another major item delivered with F.1EQ-4-200s, shown here with #4523, were huge "Irakien" drop tanks, with capacity of 2,200 litres. As usually, the plane wears no national insignia, but it has already got radio-call-sign, used during delivery flight, applied on the fin and under the cockpit.
(Dassault, via Michel Benichou)

By mid-1984, F.1EQs and F.1EQ-2s of No.79 and No.91 Squadrons, respectively were flying regular CAS-operations in the defence against almost every single Iranian offensive. Together with F.1EQ-4s from the newly established No.89 Squadron, which took over the air defence tasks from No.91, when the later began assisting No.79 Squadron in ground attack and interdiction operations.

A new and powerful weapon was added to the already available arsenal early that year. The Saudi-supplied Mk.84 bombs, equipped with Jupiter radar-fuses (acquired directly from South Africa). These fuses were set to detonate the weapon eight metres over the ground or water surface, maximising the destructive effect of the bomb. Their effectiveness was as such, that the IrAF began applying them on a range of Soviet-made bombs as well.

By 1985, all the Iraqi Mirage units were mainly tasked with air-to-ground operations. The Saddam AB-based No.89 Squadron was sole squadron reserved for air defence purposes. By that time the IrAF could count on some 40 Mirages, and these began ever intensive operations against important Iranian economic and

The first Mirage F.1EQ-5 – serial number #4560 – as seen while rolling for take-off for the first test-firing of an AM.39 Exocet, in France, 1984. Note the large centreline-pylon, required for mating the missile with aircraft. Various issues with this pylon caused several failures during initial Iraqi attempts to deploy this combination in combat. The main other outside difference when compared to earlier versions is the addition of VOR/Loc-antennas, below the top of the fin, introduced since the appearance of F.1EQ-4 version.
(Tom Cooper collection)

Iraqi Fighters, 1953-2003

F.1EQ-5 represented the "high-end" of what was available in sense of most modern electronic equipment for Iraqi Mirages in the mid-1980s. The F.1EQ-5 #4562 shows re-designed housings for advanced Sherloc RWRs on the leading edge of the fin, Corail chaff & flare dispensers on underside of wing (between the inner underwing pylon and fuselage), Remora ECM-pod (outboard underwing pylon), and Rafael TH side-borne--looking radar (SLAR) in aerodynamic housing under the fuselage. Despite this configuration, the Iraqis would also always add two Magic Missiles on wing-tips for operational missions! (Thomson CSF, Collection Raffiant, via Michel Benichou)

military installations, while providing close-air-support and battlefield interdiction as necessary during various Iranian offensives.

Through 1985, together with other IrAF assets, the No.79 and No.91 Squadrons became ever deeper involved in aerial offensive against the Khark Island. Initially, detachments operating from Wahda AB were mainly tasked with escort-missions for locally based Su-22s. Over the time, such operations became ever more sophisticated, Mirages deploying also SICOMOR and Caiman ECM-pods for stand-off jamming support. Joint operations of No.79 and No.81 Squadrons were a common occurrence.

Period III: Anti-Ship Attacks & Long-Range Raids

From late 1984, Iraqi Mirage-operations entered a new phase. When the first examples of the Mirage F.1EQ-5-200 variant, designed for anti-ship operations and use of PGMs, became operational.

The development of this variant took some time, and thus when it was originally ordered, in mid-1983, the French government agreed to lease five Dassault Super Etendard strike fighters as gap fillers until the new Mirage variant would become available. These five fighters arrived in Iraq in October 1983, and began flying combat operations in late February 1984. The Iraqis designated them as "Hunters".

Equipped with Agave radar, and armed with Aérospatiale AM.39 Exocet rockets, Super Etendards proved capable of operating by night and in adverse weather. They fired several dozens of Exocets in combat, mainly against merchant shipping along the Iranian coast of the Gulf. By the summer of 1985 only one was lost under somewhat mysterious circumstances. The Super Etendard pilots later formed the nucleus of No.81 Squadron. Their positive experiences with Super Etendards were crucial for subsequent Mirage F.1EQ-5 operations.

The first out of 20 Mirage F.1EQ-5s began arriving in Iraq during the summer of 1985. They were equipped with the much improved Cyrano-IVM radar, and had the main purpose in interdiction of shipping carrying Iranian oil exports from Khark Island down the Gulf. Like in the case of Etendards previously, their main armament was AM.39 Exocet. This weapon was in service of the IrAF since the late 1970s. It was used against Iranian shipping since the late 1980, but the first

Dassault Mirage F.1

Mirage F.1EQ-5 #4568 was the eighth F.1EQ-5 built for Iraq. It is seen here carrying the 2,200-litre, "Irakien" drop-tank, developed on Iraqi requirements and subsequently introduced as standard in the Armée de l'Air's fleet of Mirage F.1CTs. Also of interest are details of the "bolt-on" in-flight refueling (IFR) probe, which was a standard on the Iraqi Mirages since the F.1EQ-4 version. The addition of the IFR probes resulted in the full designation of these aircraft becoming F.1EQ-4-200, F.1EQ-5-200 and F.1EQ-6-200. Of interest is also that this version was still equipped with BF RWRs, obvious from their pointed housings on the leading edge, and the rear tip of the fin. (Dassault Aviation via Tom Cooper)

Showing its camouflage patterns to advantage, this F.1EQ is seen while on a training flight over north-eastern Iraq, in the mid-1980s. Note the complete absence of national markings: while usually applied once the plane was delivered to Iraq, many Mirages were left without any, or wore only parts of them. Others have got a full set of markings, including identification triangles on upper surfaces of both wings (application of these on only the left upper wing is not known). RP.35 drop tanks on this plane have been left in bare metal, but F.1EQ-4s are known to have been delivered with a batch of drop tanks painted in sand. (Ahmad Sadik collection)

successful combat deployment of the Mirage occurred only in the early 1985. By the end of the war, over 600 Exocets were fired, several No.81 Squadron pilots scoring dozens of hits.

The unit was swift in expanding its area of operations. In 1987, a formation of nine Mirages armed with two Mk.84 bombs each, carrying Remora ECM-pods as well as SICOMOR chaff & flare dispensers, and two Mirages armed with two AS.30L laser-homing air-to-ground missiles each, delivered a particularly decisive attack against the construction site of first two Iranian nuclear reactors, near Bushehr. A formation of MiG-23MLs was used for deception, dragging away a pair of defending Iranian F-14, while nine Mirages that participated in the raid were supplied with fuel in the air from nine other F.1EQ-4s "buddy-re-fuellers".

Except for Exocets, from early 1986, IrAF F.1EQ-5s were equipped to carry the French-made ATLIS I laser-marker and AS.30L laser-guided air-to-surface missiles, the first 120 of which were delivered in the same year. The IrAF ordered 568 AS.30Ls by 1989, but only 240 were delivered due to difficulties with outstanding payments. Initially used against naval targets, AS.30L was later used for attacks on various economic targets in Iran as well.

Simultaneously, the French have sold also 1,000 TMV 630 EBLIS kits for the 400kg Matra T200 laser-guided bombs. These were compatible with ATLIS 1. Such weapons were seldom used, however, mainly due to their high prices.

Through autumn 1987, the Iraqis went a step further, mating the seeker heads of Soviet-manufactured Kh-29L missiles carried by their Su-22M-4Ks, with the ATLIS pod. This combination was deployed for the first time during the final counter-offen-

sive that resulted in liberation of Faw Peninsula from the Iranian occupation, in April 1988, when Mirages designated the target for Kh-29L-dropping Sukhois, with immense success.

For such operations, the No.81 Squadron was usually deploying a pair of ATLIS-carrying F.1EQ-5s, escorted by four MiG-23s or MiG-29s, and two other Mirages equipped for stand-off jamming. Depending on the target, between two and four Sukhois were carrying Kh-29Ls.

Meanwhile, starting from mid-1986, the No.81 Squadron opened a series of very long-range raids, reaching deep over the Gulf to hit Iranian oil loading installations, and then ranging ever deeper over the Iranian soil for the same purpose.

On 12 August 1986, two Mirages hit Sirri Island, 650km away from the Iraqi coast, after being refuelled from four F.1EQ-4s, carrying Intertechnique "Buddy"-refuelling pods. A fortnight later, on 29 August, two Mirages hit Lavan Island, over 880km away from the Iraqi shores. On 25 November of the same year, two Mirages flew the longest-ever IrAF mission of the whole war, lasting 3 hours and 55 minutes, to hit Larak Island, over 1.250km away from Faw.

SAMP 400 GP-bombs were used in all of these attacks. Additional missions of similar type were flown especially in 1988, when No.81 Squadron raided the Nekka power plant and the Tehran's Rey Oil Refinery. In both cases, raiders were refuelled from other Mirages well inside the Iranian airspace, and while underway at a very low level.

When long-range raids into the lower Gulf confronted ever increasing Iranian aerial opposition, in early 1988, the IrAF was in the process of receiving the first Mirage F.1EQ-6s. According to Iranian intelligence sources, at least 20 Super 530D missiles were supplied as well. While evidence firmly supporting either the Iranian or the Iraqi version is still outstanding, it is certain that new Mirages have had several clashes with Iranian interceptors during that year, which apparently resulted with the losses of at least one F.1EQs, two F-14s, and an F-4E, between February and July 1988.

Deliveries of the last two F.1EQ-6s were stopped due to UN arms embargo, in August 1990. These were later used by the French Air Force for dissimilar air combat manoeuvring training against USN Grumman F-14 Tomcats and McDonnell F/A-18 Hornets. Iraq thus received 95 single- and 15 two-seat F.1s, out of some 129 airframes usually reported as being ordered. While Iranian claims go into six dozens, according to IrAF records, no more than 25 F.1EQs were lost in combat during the war with Iran.

By January 1991, the IrAF has had 29 Mirage F.1EQ/EQ-2/EQ-4s, and 34 F.1EQ-5/EQ-6s available. Equipped with Syrel ELINT-pods, Mirages flew over 60 reconnaissance sorties along the Saudi border between August 1990 and mid-January 1991, establishing a very good electronic order of battle of Coalition forces. During the war, only earlier variants – some examples of which were apparently upgraded to later standards over the time – were deployed in air combats against the aerial armada put up by the US-led Coalition, and four of these were shot down. 24 aircraft belonging to later variants, but including at least two two-seaters, have been evacuated to Iran, where they were later pressed into service with the local air force.

Problems with spares acquisitions saw the decline of IrAF Mirage operations after 1991. Ten years later, the last 12 operational examples were concentrated within the No.81 Squadron.

Dassault Mirage F.1

F.1EQ-6 was the last variant supplied to Iraq. Based on F.1EQ-5 and its Cyrano IV-M radar, it contained additional working modes, as moving-target indicator (MTI), thus making it capable of engaging low-flying targets in look-down/shoot-down mode over water. Camouflage pattern remained unchanged ever since appearance of F.1EQ, back in 1980.
(G. Fassari)

The Mirage F.1EQ was the workhorse during the war with Iran, even if the number of available airframes was lower than that of certain other types. It proved versatile and easy to maintain, offering excellent performance combined with superior precision, load-carrying and self-defence capabilities.

Camouflage, Markings & Serial Numbers

Upper surfaces and sides on most of IrAF Mirage F.1EQs were painted in colour called "sable" (sand) by French, with large splotches of "kaki I" (khaki green; a colour similar to that used in French Air Force in the 1940s) applied along the standard pattern for this type. On many photographs, the kaki colour appears rather like a version of dark earth, confusing a number of observers, but it was

Starting with #4571, several Mirage F.1EQ-5s were painted in gris marin foncé (dark sea grey) over, and Insignia White (FS17875) below, for operations over water. Their main purpose were attacks on merchant shipping along the Iranian coast of the Gulf. Note that all the maintenance stencilling was still in the same position like on aircraft painted in disruptive camouflage pattern.
(Dassault, via Tom Cooper)

Iraqi Fighters, 1953-2003

A very rare photograph of one of F.1EQ-6s (or F.1EQ-7s as they were called in the West) that was embargoed due to Iraqi invasion of Kuwait, in August 1990. The plane was subsequently taken over by the French Air Force, and flown in training exercises against USN F-14s and F/A-18s, in September 1990. (Photo by M. Jean-Jacques Petit, via Tom Cooper)

shining a distinct greenish touch when seen with naked eye and from short distance. Undersides were in aluminium grey.

Starting with Mirage F.1EQ-5 #4571, many EQ-5s and EQ-6s have had their upper surfaces and sides painted in gris bleu foncé II – dark blue grey (sometimes called "slate grey") – with undersides in colour similar to Insignia White (FS17875).

A Mirage F.1EQ-5 (left) and F.1EQ-6 (right) spotting camouflage pattern used for anti-shipping operations. (Left: Tom Cooper Collection; Right: Dassault, via Tom Cooper)

Dassault Mirage F.1

Table 9: Serial Numbers

Serial numbers were applied in black on the front fuselage only. Otherwise, IrAF Mirages have worn all the servicing and maintenance markings as applied on standard French Air Force aircraft of this type as well. The table below includes also observations about the fate of specific aircraft, as far as this is known.

IrAF Serial No.	Type	Delivery	Remarks
4000	Mirage F.1BQ	27 August 1980	Flown to Iran, 1991
4001	Mirage F.1BQ	11 September 1980	Flown to Iran, 1991
4002	Mirage F.1BQ	17 June 1981	Flown to Iran, 1991
4003	Mirage F.1BQ	24 July 1981	Flown to Iran, 1991
4004	Mirage F.1EQ	29 August 1980	Arrival in Iraq: 30 Jan 1981 Flown to Iran, 1991
4005	Mirage F.1EQ	29 August 1980	Last seen Iraq, 1981
4006	Mirage F.1EQ	29 August 1980	
4007	Mirage F.1EQ	29 August 1980	
4008	Mirage F.1EQ	29 August 1980	
4009	Mirage F.1EQ	29 August 1980	
4010	Mirage F.1EQ	29 August 1980	Last seen Iraq, 1981
4011	Mirage F.1EQ	29 August 1980	
4012	Mirage F.1EQ	29 August 1980	
4013	Mirage F.1EQ	29 August 1980	
4014	Mirage F.1EQ	29 August 1980	14 kill markings, captured 2003
4015	Mirage F.1EQ	29 August 1980	
4016	Mirage F.1EQ	29 August 1980	Flown to Iran, 1991
4017	Mirage F.1EQ	29 August 1980	
4018	Mirage F.1EQ	29 August 1980	
4019	Mirage F.1EQ	14 October 1981 (last)	Flown to Iran, 1991
4020	Mirage F.1EQ-2	9 April 1981	Batch 2: 16 single-seaters
4021	Mirage F.1EQ-2	9 April 1981	Flown to Iran, damaged on landing
4022	Mirage F.1EQ-2	9 April 1981	
4023	Mirage F.1EQ-2	9 April 1981	
4024	Mirage F.1EQ-2	9 April 1981	
4025	Mirage F.1EQ-2	9 April 1981	
4026	Mirage F.1EQ-2	9 April 1981	
4027	Mirage F.1EQ-2	9 April 1981	
4028	Mirage F.1EQ-2	9 April 1981	Last seen Iraq, 1984
4029	Mirage F.1EQ-2	9 April 1981	
4030	Mirage F.1EQ-2	9 April 1981	
4031	Mirage F.1EQ-2	9 April 1981	
4032	Mirage F.1EQ-2	9 April 1981	Flown to Iran, 1991
4033	Mirage F.1EQ-2	9 April 1981	
4034	Mirage F.1EQ-2	9 April 1981	
4035	Mirage F.1EQ-2	9 April 1981	
4500	Mirage F.1EQ-4	31 December 1982	Flown to Iran, 1991
4501	Mirage F.1EQ-4	31 December 1982	Intertechnique IFR-pod testing
4502	Mirage F.1EQ-4	31 December 1982	
4503	Mirage F.1EQ-4	31 December 1982	

113

Iraqi Fighters, 1953-2003

IrAF Serial No.	Type	Delivery	Remarks
4504	Mirage F.1BQ	15 April 1982	
4505	Mirage F.1BQ		
4506	Mirage F.1EQ-4		Captured 2003, taken to USA
4507	Mirage F.1EQ-4		
4508	Mirage F.1EQ-4		
4509	Mirage F.1EQ-4		
4510	Mirage F.1EQ-4		
4511	Mirage F.1EQ-4		
4512	Mirage F.1EQ-4		
4513	Mirage F.1EQ-4		
4514	Mirage F.1EQ-4		
4515	Mirage F.1EQ-4		
4516	Mirage F.1EQ-4		
4517	Mirage F.1EQ-4		Captured 2003
4518	Mirage F.1EQ-4		
4519	Mirage F.1EQ-4		
4520	Mirage F.1EQ-4		
4521	Mirage F.1EQ-4		
4522	Mirage F.1EQ-4		Last seen Iraq, 1986
4523 Y-IBLU	Mirage F.1EQ-4		
4524	Mirage F.1EQ-4		
4525	Mirage F.1EQ-4		
4526 Y-IBLU	Mirage F.1EQ-4		
4527 Y-IBLV	Mirage F.1EQ-4		
4528 Y-IBEF	Mirage F.1EQ-4		Last seen Iraq, 1986
4529	Mirage F.1EQ-4	11 July 1984 (last)	
4556	Mirage F.1BQ	14 December 1984	
4557	Mirage F.1BQ	14 December 1984	Last seen Iraq, 1986
4558	Mirage F.1BQ	14 December 1984	
4560	Mirage F.1EQ-5	19 December 1983	AM.39 testing, flown to Iran, 1991
4561	Mirage F.1EQ-5	19 December 1983	DG-camo, AS.30L testing
4562	Mirage F.1EQ-5	19 December 1983	
4563	Mirage F.1EQ-5	19 December 1983	Kh-29L testing, flown to Iran, 1991
4564	Mirage F.1EQ-5	19 December 1983	
4565	Mirage F.1EQ-5	19 December 1983	
4566 Y-IBLU	Mirage F.1EQ-5	19 December 1983	
4567	Mirage F.1EQ-5	19 December 1983	
4568	Mirage F.1EQ-5	19 December 1983	
4569	Mirage F.1EQ-5	19 December 1983	
4570	Mirage F.1EQ-5	19 December 1983	2 kill markings
4571	Mirage F.1EQ-5	19 December 1983	DG-camo, flown to Iran, 1991
4572	Mirage F.1EQ-5	19 December 1983	
4573	Mirage F.1EQ-5	19 December 1983	
4574	Mirage F.1EQ-5	19 December 1983	
4575	Mirage F.1EQ-5	19 December 1983	

Dassault Mirage F.1

IrAF Serial No.	Type	Delivery	Remarks
4576	Mirage F.1EQ-5	19 December 1983	
4577	Mirage F.1EQ-5	19 December 1983	DG-camo
4578	Mirage F.1EQ-5	19 December 1983	DG-camo
4579	Mirage F.1EQ-5	25 February 1985 (last)	
4600	Mirage F.1EQ-6	13 January 1988	
4601	Mirage F.1EQ-6	13 January 1988	
4602	Mirage F.1EQ-6	13 January 1988	DG-camo
4603	Mirage F.1EQ-6	13 January 1988	
4604	Mirage F.1EQ-6	13 January 1988	
4605	Mirage F.1EQ-6	13 January 1988	
4606	Mirage F.1EQ-6	13 January 1988	Last seen 1989, Iraq
4607	Mirage F.1EQ-6	13 January 1988	
4608	Mirage F.1EQ-6		
4609	Mirage F.1BQ	13 January 1988	
4610	Mirage F.1BQ	13 January 1988	
4611	Mirage F.1BQ	13 January 1988	Flown to Iran, 1991
4612	Mirage F.1BQ	13 January 1988	Flown to Iran, 1991
4613	Mirage F.1BQ	13 January 1988	Flown to Iran, 1991
4614	Mirage F.1BQ	13 January 1988	
4615	Mirage F.1EQ-6	Through 1988-1989	
4616	Mirage F.1EQ-6	Through 1988-1989	
4617	Mirage F.1EQ-6	Through 1988-1989	
4618	Mirage F.1EQ-6	Through 1988-1989	Captured 2003
4619	Mirage F.1EQ-6	Through 1988-1989	Flown to Iran, 1991
4620	Mirage F.1EQ-6	Through 1988-1989	Flown to Iran, 1991
4621	Mirage F.1EQ-6	Through 1988-1989	
4622	Mirage F.1EQ-6	Through 1988-1989	
4623	Mirage F.1EQ-6	Through 1988-1989	
4650	Mirage F.1EQ-6	Through 1988-1989	Flown to Iran, 1991
4651	Mirage F.1EQ-6	Through 1988-1989	SG-camo, flown to Iran, 1991
4652	Mirage F.1EQ-6	Through 1988-1989	Flown to Iran, 1991
4653	Mirage F.1EQ-6	Through 1988-1989	Flown to Iran, 1991
4654	Mirage F.1EQ-6	Through 1988-1989	Flown to Iran, 1991
4655	Mirage F.1EQ-6	Last delivery 1990	Flown to Iran, 1991
4656	Mirage F.1EQ-6		
4657	Mirage F.1EQ-6		
4658	Mirage F.1EQ-6		
4659	Mirage F.1EQ-6		
4660	Mirage F.1EQ-6		Flown to Iran, 1991
4661	Mirage F.1EQ-6		Not delivered
4662	Mirage F.1EQ-6		Not delivered
7445	Super Etendard	October 1983	Ex 64
7446	Super Etendard	October 1983	Ex 65
7447	Super Etendard	October 1983	Ex 66
7448	Super Etendard	October 1983	Ex 67
7449	Super Etendard	October 1983	Ex 68

Chapter 10

Sukhoi Su-7

Su-7BMK
(ASCC Code: Fitter)

Service History

In May 1966, an Iraqi delegation on a visit in Moscow concluded an agreement for delivery of 34 Su-7BMK fighter-bombers. Twenty additional aircraft of the same type were ordered in July 1967. In the spring of 1967, the first group of IrAF pilots was sent to the Soviet Union for training on their new mounts.

The first 18 Sukhois arrived in Iraq in October 1967, entering service with No.1 Squadron, activated by an order issued on 3 December 1967, and stationed in Kirkuk.

One year later, in October 1968, 14 additional Su-7s followed, delivered in two batches of ten and four airframes, respectively. They entered service with the

Nicely illustrating the camouflage pattern applied sometimes in the early 1970s on IrAF Su-7BMKs, this photograph shows also the most widely used weapons configuration for this type in Iraqi service. This consists of two PTB-600 drop tanks under the centerline pylon, and four UV-16-57U pods under wings. Reconstruction of this aircraft is shown on p.114.
Note the fin of an example still without camouflage colours, directly above the pilot, and shown on artwork overleaf.
(Tom Cooper Collection)

Iraqi Fighters, 1953-2003

The first batch of Su-7BMKs delivered to Iraq were all left in natural metal overall. The only areas covered with colour were the supersonic shock cone and the antenna on the top of the fin, painted in gloss green, as well as an "anti-glare" panel in front of the cockpit, painted in light grey. National markings were applied in usual places, and it seems that none were carried on upper wing surfaces. Serial numbers were applied on their usual place, on front of the fuselage, in black.

Reconstruction of the Su-7BMK from p.117 showing the aircraft in a camouflage pattern consisting of sand and olive drab on upper surfaces, and light blue on lower surfaces. This pattern was applied only after the aircraft was delivered to Iraq. Examples camouflaged before their delivery wore a pattern roughly similar to that of MiG-21MFs, with light sand and dark olive green.

re-activated No.5 Squadron, officially established on 1 February 1969, and also based in Kirkuk.

During the same year, another 20 Su-7s entered service with No.8 Squadron, a former bomber unit, previously equipped with Ilushin Il-28s, but now moved to newly-renamed Wahda AB, near Basrah.

The IrAF should thus have got a total of 54 Su-7s, including few two-seaters (the exact break-down remains unknown). Sukhoi's records differ significantly in this regards. They show deliveries of 20 Su-7s in 1968, 12 in 1969, and three at an unknown date in the late 1960s or early 1970s. Another 15 in 1973, 24 in 1974, and four others at an unknown date. This means a total of 74 airframes. The fact is that available IrAF records do not mention any new Su-7s entering service after 1969, and even less so after the 1973 War. Quite on the contrary: two out of three units equipped with the type were re-equipped with other aircraft following that war (more about this later).

118

The equipment of Iraqi Su-7BMKs was quite similar to that of aircraft delivered to other air forces, and included the simple SRD-5 Baza-6 ranging radar and Sirena-2 RWR, as well as SRO-2 Khrom Nikel IFF-transponder. Except for their standard armament of two NR-30 cannons, Iraqi Su-7s were usually armed with UV-16-57 and UV-32-57 pods for unguided rockets, as well as various general purpose bombs – foremost FAB-250M-62. All of these are known to have been used in combat, against the Israelis, and Kurds.

IrAF Su-7s were deployed in combat for the first time very shortly after the No.1 Squadron became operational, against Kurds in northern Iraq. In 1973, two units were deployed to Syria: No.1 Squadron was based at Blei airfield, while No.5 operated from Damascus IAP. Both flew hundreds of attack sorties, mainly against Israeli armoured formations on Golan, but also missions of armed reconnaissance, as well as several raids deeper behind the Israeli lines. The two units paid a dear price for their efforts, losing a total of 12 aircraft in the process.

After the war, in 1974, the No.1 Squadron was reorganized, handing over its surviving mounts to other units, and receiving Su-20s. No.5 Squadron followed that pattern, and was equipped with Su-22s in 1977. The remaining Su-7s were then operated by No.8 Squadron, meanwhile moved to Abu Ubaida AB, near al-Kut. This unit became an Operational Conversion Unit (OCU), training new pilots for Su-20s and then Su-22s. Nevertheless, it did participate in the early stages of the war with Iran, starting in September 1980, flying close air support sorties. Three of its Su-7BMKs were lost during that conflict. No.8 Squadron was disbanded in 1984, and its tasks taken over by No.44 Squadron, meanwhile equipped entirely with Su-22s.

In 1986, Egypt donated some 30 Su-7s in very poor condition to Iraq. This deal was originally to include remaining Egyptian Tupolev Tu-16 bombers as well, but Cairo bowed to immense US pressure, and would not deliver these to Iraq. The Egyptian Sukhois were in poor condition, barely more than derelicts, and all were used as decoys at various IrAF airfields.

Camouflage, Markings & Serial Numbers

Originally delivered in natural metal overall, all the IrAF Su-7s have been camouflaged by the time of the 1973 War with Israel. There have been two camouflage patterns, of which one consisted of thick, irregular stripes of olive drab over sand on upper surfaces, and light blue on lower. Exact details about the other camouflage pattern, apparently applied on aircraft of No.8 Squadron already before their delivery to Iraq, remain unknown, except that it was described as "somewhat different, more adapted to the desert environment of southern Iraq".

National markings should have been applied on usual spots, but it remains unknown whether any were carried on upper wing surfaces. Camouflaged examples appear to have had a larger fin flash than aircraft originally delivered in natural metal overall. Serial numbers were stencilled in black on front fuselage of the aircraft on their delivery. Later on, it was applied with a brush, once the aircraft were camouflaged.

Iraqi Fighters, 1953-2003

Table 10: Serial Numbers

Serial numbers were applied in black on front fuselage, roughly below the cockpit area, and this practice was continued once the aircraft were camouflaged. According to Sukhoi records, their serial numbers were as follows in the list below; sadly, there no photographs are available that would confirm or deny the Soviet records. It is worth mentioning that at least few Su-7BMKs served long enough to be re-serialled, in 1988.

IrAF Serial No.	Type	Delivery	Remarks
755	Su-7BMK	October 1967	Sukhoi records mention first delivery only in 1968
756	Su-7BMK	October 1967	
757	Su-7BMK	October 1967	
758	Su-7BMK	October 1967	
759	Su-7BMK	October 1967	
760	Su-7BMK	October 1967	
761	Su-7BMK	October 1967	
762	Su-7BMK	October 1967	
763	Su-7BMK	October 1967	
764	Su-7BMK	October 1967	
765	Su-7BMK	October 1967	
766	Su-7BMK	October 1967	
767	Su-7BMK	October 1967	
768	Su-7BMK	October 1967	
769	Su-7BMK	October 1967	
770	Su-7BMK	October 1967	
771	Su-7BMK	October 1967	
772	Su-7BMK	October 1967	
773	Su-7BMK	October 1967	
774	Su-7BMK	October 1967	
808	Su-7BMK	October 1968	Sukhoi records mention delivery in 1969
809	Su-7BMK	October 1968	
810	Su-7BMK	October 1968	
811	Su-7BMK	October 1968	
812	Su-7BMK	October 1968	
813	Su-7BMK	October 1968	
814	Su-7BMK	October 1968	
815	Su-7BMK	October 1968	
816	Su-7BMK	October 1968	
817	Su-7BMK	October 1968	
870	Su-7BMK	1969	Batch of 3 delivered on unknown date
871	Su-7BMK	1969	
881	Su-7BMK	1969	
884	Su-7BMK	1969	Batch of 14 delivered in 1973 according to Sukhoi
885	Su-7BMK	1969	

Sukhoi Su-7

IrAF Serial No.	Type	Delivery	Remarks
886	Su-7BMK	1969	
887	Su-7BMK	1969	
888	Su-7BMK	1969	
889	Su-7BMK	1969	
890	Su-7BMK	1969	
891	Su-7BMK	1969	
892	Su-7BMK	1969	
893	Su-7BMK	1969	
894	Su-7BMK	1969	
895	Su-7BMK	1969	
896	Su-7BMK	1969	
897	Su-7BMK	1969	
898	Su-7BMK	1969	
978	Su-7BMK		Bach of 24 ordered; probably never delivered
979	Su-7BMK		
980	Su-7BMK		
981	Su-7BMK		
982	Su-7BMK		
983	Su-7BMK		
984	Su-7BMK		
985	Su-7BMK		
986	Su-7BMK		
987	Su-7BMK		
988	Su-7BMK		
989	Su-7BMK		
990	Su-7BMK		
991	Su-7BMK		
992	Su-7BMK		
993	Su-7BMK		
994	Su-7BMK		
995	Su-7BMK		
996	Su-7BMK		
997	Su-7BMK		
998	Su-7BMK		
999	Su-7BMK		
1000	Su-7BMK		
1227	Su-7BMK		Batch of 24 ordered; probably never delivered
1228	Su-7BMK		
1229	Su-7BMK		
1230	Su-7BMK		

Chapter 11

Sukhoi Su-20/-22

Su-20, Su-22M, Su-22, Su-22M-3K & Su-22M-4K (ASCC Code: Fitter)

Service History

The first ten Su-20s arrived in Iraq in a particularly surprising manner, during the War with Israel, in 1973. Several Soviet Antonov An-12 transports landed at Rashid AB, in Baghdad, full with crates containing disassembled Su-20 fighter-bombers. The Iraqis loaded the crates on several tank transports and brought them to Hurrya AB, near Kirkuk.

An Iraqi delegation, including several Su-7BMK-pilots, inspected a Su-20 during a visit at Kubinka AB, in 1972. Though there was a recommendation for the type, the Iraqis issued no order. The arrival of Lyulka AL-7F-1 powered Su-20s at Rashid AB in October 1973 thus came as a surprise for them. The remaining eight planes from this order followed in 1974, and they were delivered in the usual manner – by a ship to Basrah.

An early photo of a Su-20 section belonging to No.1 Squadron IrAF, taken shortly after their delivery to Iraq. At first look, all the aircraft appear to have been painted in silver-grey overall: details of camouflage are apparent only at a closer look of original.
(Ahmad Sadik Collection)

Iraqi Fighters, 1953-2003

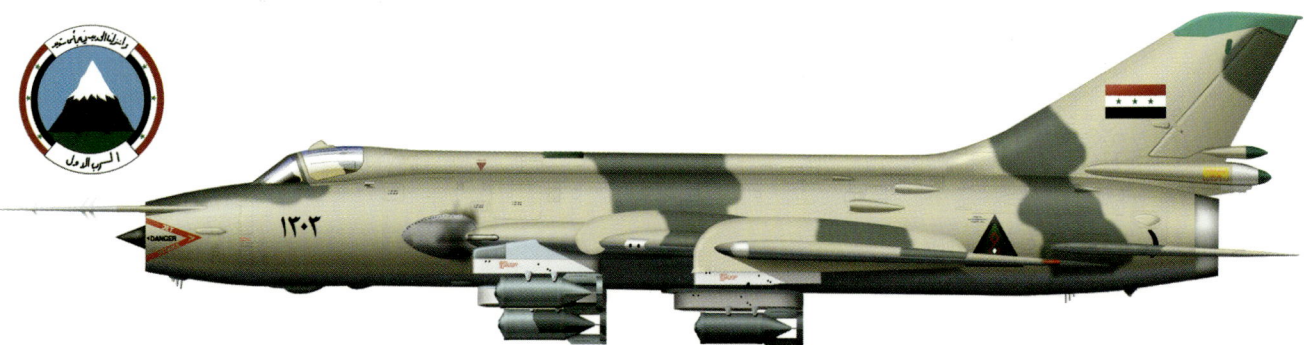

As originally delivered to IrAF, and for most of their careers in Iraq, Su-20s wore this camouflage pattern, consisting of light sand and olive drab on upper surfaces, and light blue on lower surfaces. Fin flash was applied without inclination, and triangles worn on six spots. It is interesting to note that IrAF Su-20s do not have the additional set of triangles on forward fuselage. The aircraft is shown carrying a load of six RBK-250 ZAB-2 CBUs. The insert shows the insignia of No.1 Squadron.

Su-22 #2079 as seen on photo from summer of 1980. The aircraft is shown in configuration as flown during the opening strike on Iranian airfields, on the afternoon of 22 September 1980, carrying a pair of FAB-250ShNs on inboard underwing pylons. Drop tanks carried on outboard underwing pylons were "deleted" in order to show the details behind them. Of interest is the additional set of triangles, applied below the cockpit. The plane was in service with No.44 Squadron at Hurrya AB at the time.

Reconstruction of Su-22 #2250, on the basis of photographs #1107, #1108, and #1109, showing the details of a darker version of standard camouflage pattern for this variant in IrAF service. The plane has got a second set of identification triangles below the cockpit, the light grey anti-glare panel, as well as the sign of excellence, in the form of Iraqi flag on a shield, outlined in yellow and white. It is shown in configuration for reconnaissance missions sometimes seen from autumn 1980, but especially so through 1981: carrying the massive KKR-1 reconnaissance pod, and SPS-141MVG-E ECM-pod.

Sukhoi Su-20/ -22

Once all the 18 aircraft were at al-Hurrya AB, they were put together by a group of Soviet technicians and test-flown. The type entered service with No.1 Squadron, which was to fly it until 2003.

Right from the start, Su-20s saw very intensive service with No.1 Squadron. The type experienced its baptism of fire already during the Kurdish rebellions, in northern Iraq, in 1974 and 1975. No.1 Squadron re-started sporadic operations against Kurds in the summer of 1980, and was one of main protagonists of the Operation Qadessiya – the opening IrAF raid against Iranian airfields, on the afternoon of 22 September 1980.

Since its service entry, the type became a favourite of many Iraqi pilots, and proved survivable in the face of all sorts of combat damage – ranging from hits by AAA of various calibre, up to near misses by AIM-54A Phoenix missiles. The version delivered to Iraq had four hardpoints under fuselage (compared to only two on Su-20s delivered to Egypt and Syria, barely one year earlier). Main armament deployed in combat during the first ten years of type's service in Iraq were various "iron" bombs, foremost FAB-50/-100/-250 and 500kg (including parachute-retarded versions).

After a short period during which the whole fleet was grounded due to engine problems, through winter 1980-1981, from 1982 IrAF Su-20s were often deployed for flying combat air patrols, armed with four R-3S missiles, along with two drop tanks under the centreline. For the rest of the war, No.1 Squadron flew countless close-air-support sorties mainly against Iranian units deployed along the northern frontlines.

First Su-22s were to follow barely two years after Su-20s. After carefully studying lessons from 1973 War with Israel, and Kurdish insurgency of 1974-1975, in 1976 the IrAF launched a modernisation program, aiming at expanding its ground attack capabilities. The pillar of this project became the purchase of Su-22s, put on order in large numbers within a short period of time.

The first 36 Su-22s (S-32MKs, equipped with Tumansky R-29 engine), easily recognizable by a chin fairing for Doppler speed & drift sensor, entered service with newly-established No.44 and No.109 Squadrons. Drawn to fly this new type were mainly experienced Su-7BMK-pilots, but there were also few novices, arriving straight from Flight Academy, in Tikrit. No.44 Squadron was based at Hurrya AB, in northern Iraq, and its pilots trained to fly over mountainous terrain. No.109 Squadron's base was at Wahda AB, in southern Iraq, and the pilots trained not only to operate over the flat local terrain, but also the waters of the Gulf. This decision

A reconstruction of a Su-22M or Su-22M-3K, as flown either by No.5 Squadron in late 1980 or early 1981, or No.69 Squadron in later years. This drawing is based on pieces of wreckage of several different Su-22s shot down over Iran in early and mid-1980s, and shows a camouflage pattern of dark sand and dark olive drab on upper surfaces, as well as "Russian light blue" on lower surfaces. Of interest is also the anti-glare panel in light grey — as applied on all the Iraqi Su-22M/M-3K/M-4Ks, and also Su-22s wearing the darker variant of the standard camouflage pattern. National insignia was applied on eight spots. Illustrated is the weapons configuration of six UV-32-57 rocket pods — the most massive of this kind, and rather rarely seen after 1981. More often, only four such pods were carried instead.

Iraqi Fighters, 1953-2003

The second most widespread camouflage pattern seen on Iraqi Su-22Ms, Su-22M-3Ks and –M-4Ks, consisted of dark earth brown and olive drab colours on upper surfaces, and light blue on lower surfaces. The additional set of national markings was almost always worn on front fuselage, as well as the excellence insignia (shown in detail on insert). The illustrated serial number was seen on the wreckage of a Su-22M or Su-22M-3K shot down near Abadan, in February 1986, but Su-22Ms and Su-22M-3Ks of No.44 Squadron were often deployed in illustrated configuration – carrying a hefty load of six FAB-500M-62 bombs – right up to the end of war.

Another version of camouflage pattern in dark earth brown and olive drab is shown on this reconstruction of several Su-22s shot down over Iran. The plane is shown carrying a single FAB-250M-62 bomb on inboard underwing pylon, and a MBD3-U6-68 multiple ejector rack with four FAB-100 bombs on outboard pylon for purpose of illustration only. Iraqi Su-20/22-units were always arming their aircraft with bombs of the same calibre only: mixed loads were extremely unusual. Interestingly, sometimes during the second half of the war with Iran, the practice of applying national insignia on the front fuselage as well was apparently dropped.

This Su-22M-4K is shown in a configuration that became a standard very late during the war, when this version was used for deployment of Kh-29L guided bombs in conjunction with ATLIS-carrying Mirage F.1EQ-5s. Contrary to the standard practice in most of Warsaw Pact air forces, Iraqis tended to mount the APU-58 launch rail with Kh-29L on the left outboard underwing pylon – apparently for easiness of access. As a counterweight, fully-loaded UV-32-57 rocket pods and drop tanks were carried on the other side of the aircraft. Illustrated is also the SPS-141MVG-E ECM-pod, frequently carried by the type at that time. Interestingly, this pod could not be carried when the plane was loaded with Kh-28 anti-radar missiles, then it would interfere with the Myetel targeting pod required for their deployment.

Sukhoi Su-20/ -22

These two photos nicely illustrate the standard camouflage pattern applied on IrAF Su-20s (regardless which sub-variant), the application of pre- and post-1988 serial numbers, and national markings. Note that there is no inclination of the fin flash, and that identification triangles were applied in six positions. (Above: Ahmad Sadik Collection, Left: US DoD)

prevented the two units to provide mutual support, even when there was dire need for this, as once the war with Iran erupted, major battles were fought in the south.

Equipped to quite high standards, Su-22 have not only got a Fon laser marker in intake cone, but also Kh-23 guided missiles, in addition to standard assortment of general-purpose bombs. Several airframes were compatible with KKR-1 reconnaissance pods (containing five reconnaissance cameras and extensive ELINT-gathering array) – even as the later proved difficult to operate.

Nevertheless, Iraqi pilots found the Kh-23E almost impossible to use in combat: following the launch, the pilot had to fly the aircraft with one hand, keeping the crosshairs in his sight centred on the target while guiding the missile with the other. Although the Soviets provided a simulator for training with Kh-23s, they would not deliver the semi-automatic version of the weapon, and thus the whole effort was largely in vain. The only instance the Kh-23 was used in combat occurred in 1982, when an IrAF Su-22 attempted to hit a bridge on Karoun River, in southern Khuzestan – missing completely.

Iraqi Fighters, 1953-2003

Wreck of one Iraqi Su-20 as seen after being destroyed by US Army troops on one of forward airfields in southern Iraq, in March 1991. The camouflage pattern appears to have been the darker version of the standard pattern. Of interest is the application of the triangle on lower surface of port wing. This is very close to the tip and inclined at 90 degrees to the leading edge. Obvious is also the PTB-1150 drop tank – one of the largest drop tanks used on IrAF Su-20/22s. (Tom Cooper Collection)

Together with subsequent versions, original Su-22s bore the brunt of air war against Iran, and – down to two or three examples – all were lost in combat by 1988. Wreckages of remaining planes were later found at Habbaniyah.

The third variant to reach Iraq (aside from two-seat versions), first entering service with No.5 Squadron, was Su-22M, 18 of which entered service in 1978. In order to illustrate the importance of this delivery, the Soviets deployed one of their most experienced test-pilots as instructor to Iraq.

The new version has got a terrain following radar inside the lower part of the nose, and was fully compatible with a wide range of weapons, including several types of guided missiles and bombs. At least in theory, they could be equipped with SPS-141 ECM-pods, but these became available only from autumn 1980, and entered regular service only during the following year.

In 1981, the Soviets agreed to upgrade six Iraqi Su-22Ms into Su-22M-2Ks. This work was undertaken inside hangars of IrAF Technical Wing in Hurrya AB, and was relatively simple, since aircraft were already wired for most of modifications, and there was enough space for additional avionics blocks.

Carried on APU-58 rail under the fuselage, the clearance between the Kh-28 and the ground was so small, its lower fin had to be folded before loading on aircraft: it would unfold following the missile separation from aircraft. On the other side, a recess had to be cut in the lower fuselage of the aircraft, in order to accommodate the upper fin of the missile!

Two versions of Kh-28 became available to Iraq: Kh-28E, designed to attack radar illuminators of US-made MIM-23A HAWK and MIM-23B I-HAWK SAMs, and Kh-28C for attacks on early warning radars. There was no outside difference on missiles, however, and modified Su-22M-2Ks differed only in their internal equipment. Since Kh-28 was an offensive weapon with considerable range, there was little to no requirement for carrier aircraft to penetrate enemy airspace. For this reason, as well as because an ECM-pod would likely disturb the work of Metel targeting pod, no SPS-141s were carried on operational sorties.

Over the time, sufficient numbers of additional Su-22M-2Ks were acquired not only to replace losses, but also equip the entire No.5 Squadron with them. In turn, older Su-22Ms were transferred to No.44 Squadron, which assumed the role of an OCU. Final batch of 18 Su-22M-2K to arrive was already equipped with the same nav/attack system like original Su-17 (minus capability to deploy nuclear weapons), and was for a short period of time the most capable fighter-bomber in Iraqi arsenal. They were used to re-equip No.109 Squadron.

Another IrAF Su-20, found abandoned in a field Kirkuk, in 2003. The serial number is #20520 — one of the highest known in this sequence and for that variant. The plane also shows a slightly altered version of the "darker" camouflage pattern applied to this variant, with the border between light earth and light blue colour drawn down the middle of the fuselage. (US DoD)

In the early 1984, the IrAF received 18 Su-22M-3Ks (S-52MKs). Outfitted with Klyon laser rangefinder and target designator, and ten hardpoints for carriage of ordnance and equipment, this version was a leap forward in terms of avionics and weaponry. It entered service with newly-established No.69 Squadron, based at al-Bakr AB. Su-22M-3K was the first variant compatible with the full array of Soviet-made precision-guided-munition (PGM), and it arrived together with significant consignment of Kh-25ML semi-active laser-homing-, Kh-25MP anti-radar- and radio-guided Kh-25MR missiles, as well as Kh-29L laser-homing missiles. Besides, it could carry R-60 missiles for self-defence, though these were never carried. No.69 experienced its baptism of fire on 15 August 1985, in a mission led by Squadron CO himself, that proved the versatility of the new version. Each aircraft was equipped with four FAB-500ShN parachute-retarded bombs under the fuselage. Also carried were two 800-litre drop tanks on outboard underwing pylons, one SPS-141MVG on left inboard pylon and an UV-32-57 rocket pod on right inboard pylon (just to balance the aircraft). KDS-23 chaff & flare dispensers were carried as well. Despite its success in combat, no additional Su-22M-3Ks were delivered ever again. Nevertheless, the version participated in all battles from 1986 until the end of the war, flying close-air-support, battlefield interdiction, interdiction and electronic support missions, as well as raiding vital Iranian economic installations.

The reason for Iraqis making little use of Kh-29L missiles was its problematic combat deployment. The problem was that following release of the missile, the aircraft had to continue flying towards target and could not manoeuvre to evade anti-aircraft defences. With sufficient numbers of AS.30Ls at hand, the Iraqis preferred the later weapon for their operations.

The final Su-22 version to enter service was Su-22M-4K, 36 of which were delivered between 1986 and 1987, to enter service with No.5 and No.109 Squadrons. Aside from being equipped with a fixed shock cone with new Klyon-54 laser marker, and lower-capacity fuel tanks, Su-22M-4K has also got Orbita-20-22 computer, which integrated the work of all the on-board navigation and attack equipment. The other addition was compatibility with Kh-29T electro-optically guided missiles.

The arrival of Su-22M-4Ks resulted in small reorganisation of the IrAF Su-22 fleet. While most of older Su-22Ms and Su-22M-2Ks were transferred to No.44 OCU, No.5 Squadron retained four Su-22M-2Ks for SEAD purposes, until the stocks of Kh-28 missiles were almost exhausted, in 1988. They were then replaced by Kh-25MP-toothing Su-22M-4Ks.

Iraqi Fighters, 1953-2003

Scene from Hurrya AB in summer of 1980, showing three Su-22s of No.44 and No.5 Squadron, #2079 in the foreground and #2077 in the centre. Sadly, the serial number of the third plane — obviously painted in the darker version of the standard camouflage pattern — is out of sight. Of interest is the full set of markings, as well as warning stencilling in English. (Ahmad Sadik Collection)

Close-up of the starboard and port side of front fuselage belonging to one of reportedly only two IrAF Su-22s that survived the war with Iran (though their serial numbers indicate that few additional examples remained intact by 1988 as well): #22588, as found in Habbaniya, in 2003. Of interest are details like additional identification triangle, and stencils in English, as well as the rest of its old serial number -2076. The picture below shows the total of starboard side of the same plane, with details of camouflage. (Tom Cooper Collection)

130

Sukhoi Su-20/ -22

Port view of Su-22 #22596, as seen on the scrap yard next to the western runway of Habbaniya AB, in early 2006. Note the additional set of identification triangles, and remnants of both, the pre- and post-1988 serial numbers on front fuselage. (Tom Cooper Collection)

Wreckage of an Iraqi Su-20, as put on display in Iran, several times. It shows the remnants of original standard pattern for this version. Sadly, neither the serial number nor the construction number are visible. The national insignia on port side has been almost completely washed away. The plane appears not to have been shot down, or to have crashed in Iran, but rather captured on one of Iraqi airfields during the uprising of April 1991, and taken to Iran. (Farzad Bishop)

It is interesting to observe, that Kh-25MP was relatively seldom used against Iran. The first round was launched in combat already on 17 April 1986, to score a direct hit on the illuminator antenna of an Iranian HAWK SAM-site, but subsequently only few rounds were expanded, as the threat from Iranian air defences in the areas where these aircraft usually operated diminished significantly towards the end of war.

Meanwhile, the Iraqis worked intensively on solving the problem of Kh-29L's deployment. By September 1987, they managed to mate its homing head with French-made ATLIS designator, carried by Mirages. The result was tactical cooperation of very special kind: a Mirage equipped with French-made laser designator, would circle the target and designate for two Sukhois and their Soviet-made laser-guided-bombs. Between 17 April 1988 and 23 July 1988, several dozens of such operations were launched, with particularly devastating results – especially since the Kh-29L packed a significant punch and penetrating capability.

Carrying the full burden of war with Iran, Iraqi Su-20s/-22s/-22Ms/-22M-2Ks/-22M-3Ks and Su-22M-4Ks units suffered a loss of 64 fighters in combat. Nevertheless, they remained very popular with all associated with it. The IrAF, its pilots, technical personnel, and especially the high command. Renowned for stability in flight and combat, simple maintenance and sturdiness, good payload, capability to survive punishment and return the pilot safely back to base, they were used for all but tasks that required long endurance. The IrAF therefore continued purchasing additional aircraft right until the end of war, not only to replace losses, but also expand the fleet and establish additional units.

In fact, starting from 1986, ever larger formations of Iraqi Su-22s operated over the frontlines, regularly saturating Iranian air defences to the point where these could not offer serious resistance.

Following the war, the IrAF Su-22-fleet has had very little respite before the next major conflict. By July 1990, an additional squadron was deployed to Wahda AB,

Iraqi Fighters, 1953-2003

An early Su-22M, apparently wearing the serial number 2026, as seen shortly before the outbreak of war with Iran. Note also the insignia of excellence to the right of the serial number, applied on almost all the Iraqi Su-22Ms: this insignia indicated also the importance of Su-22 for IrAF. Further of interest are traces of heavy use of the cannon, as can be seen by darkened spots over the gun blast panel (made of steel). The camouflage pattern probably consisted of dark earth brown and olive drab on upper surfaces.
(Ahmad Sadik Collection)

reinforcing the locally-based No.109. Both were soon to find themselves embroiled in the swift invasion of Kuwait, beginning on early morning of 2 August 1990. The opening raid of this war caught the Kuwaitis with unprepared. The only defence encountered by the first four of No.109 Squadron's Su-22M-4Ks was a sole Kuwaiti MIM-23B I-HAWK SAM-site on Bubiyan Island, at the time under command of a US instructor. Although the Kuwaitis later claimed that their SAMs have shot down 14 Iraqi aircraft, actually hit was one Sukhoi, and a MiG-23BN from a formation that took off from Ali Ibn Abu Talib AB. Both pilots perished, and no trace was found of them ever again. In response, one of the Sukhois fired a single Kh-25MP anti-radar missile, forcing the I-HAWK site to shut down its radar.

During the autumn 1990, southern IrAF Su-22-squadrons were prepared for attacks against Coalition forces in Saudi Arabia. With help of information gathered from ELINT-operations conducted by Mirage squadron, plans were developed for attacks on early warning ground radars with help of Kh-28C missiles (stocks of Kh-28Es were exhausted by the time). The IrAF also experimented with its Sukhois, and in one instance a Su-22M-4K obtained a lock-on with Kh-28C on APY-1 radar of an USAF AWACS underway inside the Saudi airspace, over 250km away.

In January and February 1991, IrAF Su-20/-22s were heavily hit by US-led Coalition onslaught. At least 14 aircraft of various versions have been destroyed on the ground. In the end the IrAF decided to reach back on an agreement with Iran, from previous summer, and a total of 40 Su-22M/M-3Ks and M-4Ks, as well as four Su-20s, were evacuated to the safety of Iranian airfields during February 1991. Two additional aircraft have been shot down while attempting to do so, while at least one Su-20, and a Su-22UM-3K, crashed on Iranian airfields due to fuel starvation, and one or two other Su-20s have been captured by pro-Iranian insurgents during the following uprisings, in April 1991, and taken to Iran as well. Two unarmed Su-22s were shot down by USAF F-15s, while on transfer flight from Hurrya to Tikrit, following the war.

What was left of IrAF "Fitter"-fleet was languishing for most of the 1990s. In 1992, the surviving aircraft were reorganised into three units, but additional airframes were subsequently lost in various accidents. By 1998, some of Su-22M-4Ks

Sukhoi Su-20/ -22

Wing of an IrAF Su-22M, as found on the scrap yard of Habbaniyah AB, in early 2006. It shows the traces of the fourth version of camouflage pattern applied on this version, consisting of light sand, olive drab and dark earth brown. Sadly, the few details known about this pattern permit no useful reconstruction in form of an artwork. Of special interest is the position of identification triangle – pointing straight forwards when the wing was in fully swept position. (Tom Cooper Collection)

have had their RWRs and chaff & flare dispensers stripped down for use on MiG-23MLs. The year 2003, for example, has found all the remaining Su-20s of No.1 Squadron in open storage at Kirkuk, while most of Su-22s were hidden under camouflage nets, or even buried under ground around various airfields in central and northern Iraq. During that war, USAF F-16 pilots are known to have claimed at least one of Iraqi Fitters as destroyed on the ground, using a concrete-filled LGB to hit the plane parked inside an urban area.

Camouflage, Markings & Serial Numbers

Iraqi Su-20s and Su-22s wore a wide range of very different camouflage patterns. The first batch is said to have arrived painted in silver-grey overall, but already the second batch (at least the aircraft after the serial number 1300) was shipped to Iraq wearing standard camouflage pattern for export Su-20s, in light sand and olive drab, similar in form to that of contemporary MiG-21MFs and Su-7BMKs. Apparently following their first complex overhaul, all the Su-20s have got a camouflage pattern of light earth and dark olive green on upper surfaces instead, with light blue-grey on lower surfaces.

Over the time, four distinct camouflage patterns emerged. The two above-mentioned ones applied foremost on Su-20s, and a similar pattern applied on Su-22s, and another pattern consisting of dark earth and olive green on upper surfaces, applied on most of Su-22Ms, -22M-3Ks and -22M-4Ks. Finally a whole range of various irregular patterns, mainly based on sand and dark olive green, sometimes with some dark earth colour in addition, were observed on Su-22s. Of interest is to note that portions of wings usually hidden within the wing-gloves when swept fully back, were not camouflaged, but either painted in black or dark grey, or left in bare metal.

Fin flashes were applied in standard, "inclined" style, while triangles were worn not only on six, but usually on eight positions, or on four positions on fuselage only. Namely, in addition to upper wing surfaces, many IrAF Su-20s, Su-22Ms and other variants, with exception of Su-22M-4Ks, have got an additional set of triangles on the front fuselage as well. The exact background behind this measure remains unknown.

An IrAF Su-22M-4K, as found by US troops inside a HAS at Ali Ibn Abu Talib AB, in March 1991. The photograph shows foremost the details of camouflage pattern around the wing and on various pylons, as well as a Kh-28 missile (left foreground), and the corresponding launch rail below the centreline of the aircraft. The Kh-28 missile saw extensive service with IrAF in the war against Iran, when 104 rounds have been fired. (US DoD)

133

Iraqi Fighters, 1953-2003

A classic shot of Su-22M-4K #22564 put on a display on a defence exhibition in Baghdad, in May 1989. While the faded colours of this photo are misleading, the aircraft was actually painted in the most widespread camouflage pattern for this and other Su-22M-variants in Iraqi service, consisting of dark earth brown and olive drab on upper surfaces. In addition, it shows the application of ASO-2 chaff & flare dispensers.
(Ahmad Sadik Collection)

Table 11: Serial Numbers

Serial numbers were applied in black on the front fuselage, and sometimes (very rarely) repeated on wing fences, or even the fin. During the reorganisation of 1988, old serial numbers were usually simply painted over with sand or green, and new serial numbers applied over them. In order to complete the presentation, known serial numbers of two-seaters are given as well.

IrAF Serial No.	Type	Delivery	Remarks
Pre-1988 Serial Numbers			
1162	Su-20	1973	1st batch
1163	Su-20	1973	
1164	Su-20	1973	
1166	Su-20	1973	
1167	Su-20	1973	
1168	Su-20	1973	
1170	Su-20	1973	
1173	Su-20	1973	
1295	Su-20	1973	
1297	Su-20	1973	
1298	Su-20	1973	
1300	Su-20	1973	
1303	Su-20	1973	
1315	Su-20	1973	
1356	Su-20	1973	
1374 or 1574	Su-22	1976	

Sukhoi Su-20/-22

IrAF Serial No.	Type	Delivery	Remarks
1702	Su-22UM-3K		
1736	Su-22M or -22M-3K		Shot down over Khuzestan, mid-1980s
1584	Su-22M or -22M-3K		Shot down near Basrah, February 1986
1836	Su-22M or -22M-3K		
2021	Su-22M or -22M-3K		
2026?	Su-22M		Shot down near Abadan, February 1986
2050	Su-20		Later 20503
2051	Su-20		Later 22511
2076	Su-22		Later 22588
2077	Su-22		No.44 Sqn, fate unknown
2079	Su-22		No.44 Sqn, fate unknown
2157	Su-22		
2250	Su-22		No.44 Sqn, fate unknown
2574	Su-22M or 22M-3K		Shot down over Iran, February 1986
Post-1988 Serial Numbers			
20501	Su-20		Captured 2003, former 2050
20502	Su-20		Captured 2003
20503	Su-20		Captured 2003
20505	Su-20		Captured 2003
20507	Su-20		Captured 2003
20508	Su-20		Flown to Iran, damaged on landing
20511	Su-20		Captured 2003, former 2051
20512	Su-20		Flown to Iran, 1991
20513	Su-20		Flown to Iran, 1991, damaged on landing
20514	Su-20		Flown to Iran, 1991, damaged on landing
20520	Su-20		Captured 2003
20525	Su-20		Captured 2003
205??	Su-20		Destroyed 1991, A.I.A. Talib AB
22500 or -506	Su-22UM-2K		Captured 2003
22501	Su-22UM-2K		Flown to Iran, 1991
22504	Su-22UM-2K		Fate unknown
22505	Su-22UM-2K		Flown to Iran, 1991
22507	Su-22UM-2K		Flown to Iran, 1991
22509	Su-22UM-2K		Flown to Iran, 1991
22511	Su-22UM-2K		Captured 2003, Tammuz
22513	Su-22UM-2K		Flown to Iran, 1991
22515	Su-22UM-2K		Flown to Iran, 1991
22521	Su-22UM-3K		Flown to Iran, 1991
22530	Su-22UM-4K		Flown to Iran, 1991
22531	Su-22UM-4K		Flown to Iran, 1991

IrAF Serial No.	Type	Delivery	Remarks
22532	Su-22UM-4K		Captured 1991, brought to Iran; c/n 1753237204
22533	Su-22UM-4K		Flown to Iran, 1991
22539	Su-22UM-4K		Flown to Iran, 1991
22539	Su-22UM-4K		Flown to Iran, 1991
22540	Su-22M-4K		Flown to Iran, 1991
22541	Su-22M-4K		Flown to Iran, 1991
22545	Su-22M-4K		Flown to Iran, 1991
22546	Su-22M-4K		Flown to Iran, 1991, damaged on landing
22549	Su-22M-4K		Flown to Iran, 1991
22554	Su-22M-4K		Flown to Iran, 1991, damaged on landing
22555	Su-22M-4K		Flown to Iran, 1991
22557	Su-22M-4K		Flown to Iran, 1991
22558	Su-22M-4K		Flown to Iran, 1991
22560	Su-22M-4K		Flown to Iran, 1991
22561	Su-22M-4K		Flown to Iran, 1991
22564	Su-22M-4K		Sighted 1989, fate unknown
22566	Su-22M-4K		Flown to Iran, 1991
225?1	Su-22M-4K		Flown to Iran, 1991
225?9	Su-22M-4K		Flown to Iran, 1991, at Nojeh AB as of 2003
22588	Su-22		Wreck at Habbaniyah, 2003, former 2076
22596	Su-22		Wreck at Habbaniyah, 2003
22607	Su-22M-2K		Flown to Iran, 1991
22615	Su-22M-3K		Captured 2003, Tammuz
22619	Su-22M-2K		Flown to Iran, 1991
22620	Su-22M-3K		Flown to Iran, 1991, damaged on landing
22634	Su-22M-2K		Flown to Iran, 1991
22635	Su-22M-3K		Flown to Iran, 1991, damaged on landing
22654	Su-22M-3K		Flown to Iran, 1991
22658	Su-22M-3K		Flown to Iran, 1991
22659	Su-22M-3K		Flown to Iran, 1991
22660	Su-22M-3K		Flown to Iran, 1991
22661	Su-22M-3K		Flown to Iran, 1991
22663	Su-22M-3K		Flown to Iran, 1991, damaged on landing
22670	Su-22M-3K		Flown to Iran, 1991
22671	Su-22M-3K		Flown to Iran, 1991
22672	Su-22M-3K		Flown to Iran, 1991

Chapter 12

Sukhoi Su-24

Su-24MK
(ASCC Code: Fencer)

Service History

The Soviets attempted to sell the Su-24 to Iraq for the first time already in 1986, when a delegation of some 100 pilots and technicians arrived at H-3, together with two Su-24MKs. The plane was test-flown by two Iraqi pilots, but no order issued immediately. The later followed only in 1988, requesting delivery of 18 aircraft. These arrived all in the early summer of 1989, entering service with re-established No.8 Squadron, based at al-Bakr AB. Most of pilots selected to fly the new type were former Su-22-pilots. Together with aircraft, the Iraqis purchased also comprehensive maintenance facilities. Tthese were also established at al-Bakr.

The arrival of Su-24 Fencers in Iraq resulted in the IrAF retiring its remaining Tupolev Tu-16s and Tu-22s, by the end of the same year. The only other bomber that remained were Chinese-made B-6Ds, flown by No.10 Squadron, equipped with C.611 anti-ship missiles (after the original 100 C.601s were all spent during the final year of war with Iran).

By the times of Iraqi invasion of Kuwait, in early August 1990, when an UN-imposed embargo stopped further deliveries, also a part of the second batch arrived, increasing the number of aircraft to 25. Original plans for establishing a second unit were shelved, and instead the IrAF concentrated on increasing the

IrAF Su-24MKs #24624 and #24651, as seen over Baghdad during a parade in 1989. Note the inconsistence in camouflage pattern on starboard side of these two aircraft. As no photograph showing the top of any Iraqi Su-24MK is available, it can only be guessed what was the camouflage pattern applied there looking like, or about differences between various aircraft.
(Iraqi National TV)

Iraqi Fighters, 1953-2003

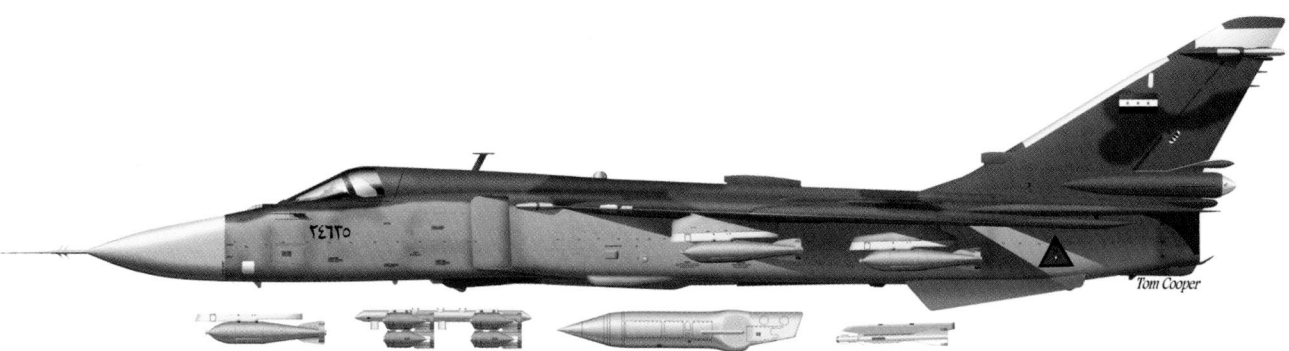

IrAF Su-24MK #24635 shown with a selection of weapons and equipment supplied for that type to Iraq. This included (from left to right); the FAB-250M-62 general-purpose bomb on BDZ-U adapter, six OFAB-100-120 bombs on MBDZ-UB multiple-ejector rack, UPAZ ("Sakhalin") refuelling pod, R-60 air-to-air missile on APU-60 launch rail. This Su-24 was the only example not flown to Iran in 1991: nicknamed "Waheeda" ("The Lonely" in Arabic), it remained in service until 2003.

combat readiness of the No.8 Squadron, planning to deploy the available Su-24MKs in the case of a war with USA or Israel. By December 1990, a plan for a major massive strike against the Israeli nuclear- and industrial infrastructure was developed. It called for 24 "clean" Fencers to deploy from al-Bakr to H-3/al-Wallid AB, be refuelled and armed (with FAB-500 "iron" bombs, but also Kh-29Ls) and then launch towards Dimona and other targets in Israel. A series of comprehensive training sessions was undertaken in late 1990s, Su-24-crews flying very low (in fact so low, they earned the type a nickname of "Dustblower" in IrAF) and learning how to avoid IrAF MiG-29s and Mirage F.1EQs.

During the first night of the following war, on 17 January 1991, the US Air Force flew a series of particularly heavy raids against al-Bakr and al-Walid air bases, causing lots of damage. Though not a single Su-24 was destroyed on the ground, the plan for attack on Israel had to be shelved. By 24 January, the situation deteriorated to a point where the IrAF decided to act according to pre-war plans, and evacuate a sizeable part of its fighter-bomber fleet to Iran. The transfer of aircraft began almost immediately, and it was during one of these flights that the – meanwhile quite famous – encounter between an Iraqi Su-24 and a pair of USAF F-15C Eagle fighter-interceptors occurred. Underway at a very low level, the Iraqi crew deployed very effective electronic countermeasures. Though the Eagles managed to establish a lock-on and fire three AIM-7M Sparrow Missiles, the Missile Approach and Warning System of the big Sukhoi fighter activated timely the chaff & flare dispensers in time, and these decoyed all the missiles. One of the US pilots managed to approach within the Sidewinder range and began firing again, launching four missiles in the process. Once again, all of these were decoyed by flares.

In total, the IrAF managed to transfer 24 Su-24s to Iran. The last aircraft, serial number 24635, had to abort the mission due to a mechanical breakdown shortly after take off. The pilot managed a safe landing and the plane survived the war.

This sole Fencer that remained in IrAF service until 2003, earned itself the nickname "Waheeda" ("The Lonely" in Arabic – and also a nickname of a famous Iraqi singer). It was captured by US troops in April 2003, and most likely transferred to the USA a month or so later, together with a sizeable batch of Mirages, MiG-25s and some other types.

Sukhoi Su-24

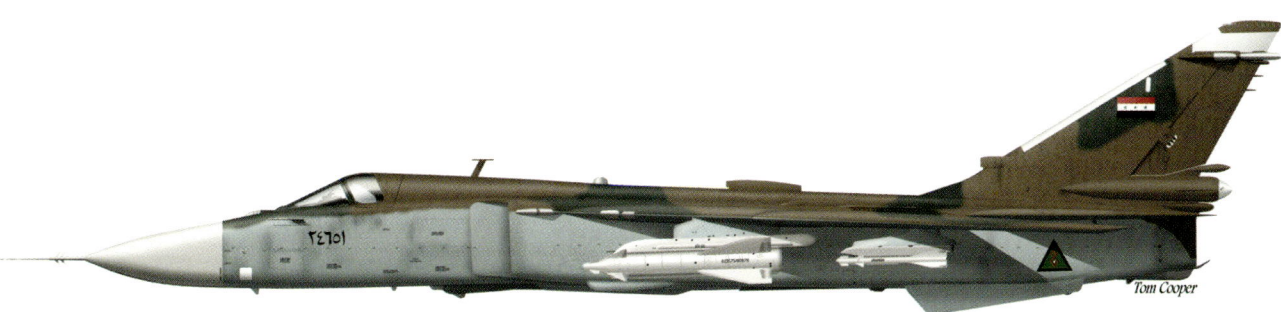

IrAF Su-24MK #24651 showing one of two configurations proposed for the mass-attack against Israeli airfields, planned for January 1991. This included a pair of Kh-29L air-to-surface missiles, and R-60s. Iraqi Su-24MKs wore the dark variant of standard camouflage pattern for export Fencers, in dark earth and olive drab on upper surfaces, as well as light blue-grey on sides and lower surfaces. Radome and all dielectric panels were painted white, and tended to become dirty within a short period of time. National markings were of similar size as applied on MiG-29s at around the same time: it is rather the size of Su-24 that made them appear "small".

Another weapon previously unknown of being sold to Iraq was Kh-58E, an export variant of this successful anti-radar missile. According to sources within the IRIAF (which obtained several Kh-58Es), the rounds supplied to Iraq were rigged only for operations against US-made MIM-23 HAWK SAMs. For deployment of Kh-58s, Fencers have to carry a LO-81 Fantasmagoria-B pod, equipped with sensors for detection of hostile emitters.

Camouflage, Markings & Serial Numbers

Iraqi Su-24s wore a version of what was apparently a standardised camouflage pattern for all exported Su-24MKs. This consisted of tan and olive drab on upper sides, and colour known as Russian light blue on sides and bottom surfaces. All dielectric panels, including the radome, were painted in white overall. Serial numbers were applied in black below the cockpit, and national insignia on at least two spots, on rear fuselage. Fin flashes were applied on usual spots. No special markings are known to have been applied during the short service the type saw with the IrAF.

Table 12: Serial Numbers

Serial numbers were applied in black on the front of the fuselage. Only four of these are known, as follows:

IrAF Serial No.	Type	Delivery	Remarks
24624	Su-24MK	1989	Flown to Iran, 1991
24625	Su-24MK	1989	Flown to Iran, 1991
24626	Su-24MK	1989	Flown to Iran, 1991
24627	Su-24MK	1989	Flown to Iran, 1991
24628	Su-24MK	1989	Flown to Iran, 1991
24629	Su-24MK	1989	Flown to Iran, 1991
24630	Su-24MK	1989	Flown to Iran, 1991
24631	Su-24MK	1989	Flown to Iran, 1991
24632	Su-24MK	1989	Flown to Iran, 1991
24633	Su-24MK	1989	Flown to Iran, 1991
24634	Su-24MK	1989	Flown to Iran, 1991
24635	Su-24MK	1990	Captured 2003, taken to USA
24636	Su-24MK	1989	Flown to Iran, 1991
24639	Su-24MK	1989	Flown to Iran, 1991
24640	Su-24MK	1989	Flown to Iran, 1991
24641	Su-24MK	1989	Flown to Iran, 1991
24642	Su-24MK	1989	Flown to Iran, 1991
24644	Su-24MK	1989	Flown to Iran, 1991
24645	Su-24MK	1989	Flown to Iran, 1991
24646	Su-24MK	1990	Flown to Iran, 1991
24647	Su-24MK	1990	Flown to Iran, 1991
24648	Su-24MK	1990	Flown to Iran, 1991, damaged on landing
24649	Su-24MK	1990	Flown to Iran, 1991
24651	Su-24MK	1990	Flown to Iran, 1991
24652	Su-24MK	1990	Flown to Iran, 1991, damaged on landing

Chapter 13

Sukhoi Su-25

Su-25K
(ASCC Code: Frogfoot)

Service History

Iraq was the first export customer of the Su-25K outside the Warsaw Pact. The first 18 Su-25Ks began arriving already in late 1985, as usual, packed in crates. They entered service with a newly established unit, the No.115 Squadron, first introduced in combat in late February 1986. An 18 additional Sukhois were delivered by early 1986, No.116 Squadron was established as well. Although for most of the war with Iran they were to operate from various forward airfields along the frontlines, especially Ali Ibn Abu Talib, Abu Ubaida, Jaliba and Artawi airfields, and Wahda AB, the main base of the IrAF Su-25K-fleet was always Tammuz. This huge airfield remained not only the headquarters of these two units, but also their technical- and

Three Iraqi Su-25s — including one of very few Su-25UBKs — seen over flying Baghdad during a military parade in 1989. (Iraqi National TV, via Ahmad Sadik)

Iraqi Fighters, 1953-2003

Reconstruction of Su-25K #25571, the Iraqi Frogfoot with lowest known serial number. It belonged to the first batch of the type delivered to Iraq, and was camouflaged in dark sand, olive drab, and chocolate brown on upper surfaces, as well as light blue on lower surfaces. This plane was found at H-3, right beside several MiG-23UBs and MiG-21s. This airfield was abandoned by IrAF in September 1996, when the USA shifted the so-called "Southern No-Fly Zone" from the 32nd to 33rd Parallel. All the operational aircraft were then taken away, while irreparable ones were destroyed, or left in place.

Su-25K #25616, as found by US troops at Tammuz AB, in 2003. Aside from the third version of camouflage pattern applied on Iraqi Su-25Ks, insert shows the "Dome of the Rock" marking, applied in commemoration for participation in the military parade that took place in Baghdad on 31 December 2001. Number 25616 was the mount of formation-leader on that occasion. The marking shows the Dome of the Rock, in gold and light blue. Also presented on this artwork are the three types of weapons and a drop tank most frequently used on IrAF Su-25Ks (from left to right): UV-32-57 pod for unguided rockets calibre 57mm; KMGU-2 cluster bomb unit of Soviet origin; Cardoen CBU-500 cluster bomb unit of Chilean origin; and a PTB-800 drop tank.

logistical base right until the very end, in 2003. Contrary to earlier times, all Iraqi Su-25K-pilots were trained by Soviets in Iraq, their instructors later complaining that they were reluctant to fly low and lacked confidence, and therefore had to be trained in medium to high altitude tactics. The fact was that the IrAF has abandoned low-level operations by the time, and mainly flew at altitudes between 3,000 and 7,000m.

All the fliers selected for Su-25-programme were combat-proven veterans from Su-22-units, and their experience showed that flying at low level over the battlefield was exposing them to the full range of Iranian anti-aircraft weapons. They

Sukhoi Su-25

Reconstruction of Su-25K #25609, with the presented camouflage being a reconstruction of what can be seen on all the photos of this aircraft's wreckage taken when it was last sighted at H-3 AB. Of interest is this entirely different camouflage pattern than known to have been applied on the first batch of Iraqi Su-25s. The serial number, though, is same in size and style, and applied in the same place as on #25571.

proved their, and the worth of their new mounts while fighting the Iranians in the spring of 1987, near Basrah, in an operation best-known as "The Great Harvest" to the Iraqis. From those times onwards, Su-25 proved the most effective asset for close-air-support in Iraqi hands.

The third batch of 18 Su-25Ks was supplied in late 1987, and entered service with No.114 Squadron. This unit was late to participate in the war with Iran, and began combat operations only in April or May of 1988.

In total, the IrAF thus received 54 Su-25Ks. Very few Su-25UBKs have been purchased, as conversion was relatively simple for any Su-22-pilots. The main armament of Iraqi Su-25Ks were Soviet-made KMGU and Chilean CB-250 cluster bombs, as well as UV-32-57 pods for unguided rockets. The IrAF never armed its Su-25s (nor any other type) with mixed loads, and carriage of R-60 air-to-air missiles occurred rather seldom.

The type saw a comparatively short service with IrAF. Though only two or three aircraft were lost during the war with Iran, at least one was destroyed on the ground, two were shot down and seven flown to Iran, in 1991. The IrAF kept the three units operational on the type for the next several years, until the lack of spares forced it to disband the No.115 and No.114 Squadrons.

Camouflage, Markings & Serial Numbers

Iraqi Su-25Ks wore three distinctive camouflage patterns, depending on their delivery date. The first batch, aircraft operated by No.115 Squadron, were painted in dark sand, olive drab, and chocolate brown over, and "Russian light blue" under. The border of the later colour was roughly along the lower third down the fuselage. Aircraft operated by No.116 Squadron have got a slightly lighter sand colour; olive drab and earth brown were applied in a different pattern, while the borders of light blue went high up, almost two thirds of the fuselage. The final batch, operated by No.114 Squadron, was painted in light blue and earth brown only, and the upper border of the light blue colour was very low, going along the lower edge of fuselage sides.

Iraqi Fighters, 1953-2003

View of the starboard side of fin on Su-25K #25571, reveals the details of national markings, and camouflage in this area. The dark olive green was apparently of such a poor quality, that it was almost completely washed away by sun, sand and rain.
(Tom Cooper Collection)

Fuselage of Su-25K #25571, as found at H-3 AB. Wings, all the avionics and most of the wiring have been taken away. The aircraft shows traces of the original camouflage pattern applied on the first two batches of Su-25s, with green surfaces on the fin and the spin above the left intake. Note the traditional inclination of national markings, which were always applied in line with the ground surface.
(Tom Cooper Collection)

View at the starboard side, of centre section, and front fuselage of Su-25K #25571, revealing not only additional details of the camouflage pattern, but also a lot about aircraft's construction.
(Tom Cooper Collection)

Detail view of #25571's cockpit section, including the serial number. It is obvious that the old serial number, from before 1988, has been painted over by green and the new one applied with help of stencils. This aircraft is reconstructed on top of p.138.
(Tom Cooper Collection)

For comparison to the left picture, here a view at the same section of Su-25K #25609, showing that the light blue colour went much higher up the fuselage. This aircraft can be seen on top of p.139.
(Tom Cooper Collection)

Sukhoi Su-25

A total view of what was left of Su-25K #25609, permitting looks at the port side of centre and rear fuselage, fin, as well as the starboard side of cockpit section. Note that there are only very few traces of green colour on this aircraft, as well as that what can be made out indicates that dark olive green was applied in different positions in the cockpit area, than this was the case with aircraft from the initial batch supplied to Iraq.
(Tom Cooper Collection)

Total view of Su-25K #25616, as it was last seen (outside one of US bases next to the former Tammuz AB, in western central Iraq), providing a good look at the overall camouflage pattern. This plane belonged to the third and last batch of Su-25s delivered to Iraq, with camouflage consisting of sand and dark brown only.
(US DoD)

Side view of #25616 as it was found at Tammuz AB, in May 2003. Of special interest is the commemorative marking described in detail on second artwork at p.138. (US DoD)

145

Iraqi Fighters, 1953-2003

Wreckage of front fuselage of an unidentified Su-25K, showing traces of dark brown colour applied on planes from the first batch, as well as the "Russian light blue" colour applied on lower surfaces. (Tom Cooper Collection)

Table 13: Serial Numbers

Only one serial of Iraqi Su-25Ks from the times before the whole fleet had its serial numbers reassigned is known. It apparently belonged to the third batch, operated by No.114 Squadron. All the other known serial numbers are from the times after the reorganisation, in 1988. In all cases, they were applied in black on the front fuselage. Three construction numbers of Iraqi Su-25Ks, all belonging to aircraft destroyed by US troops at Ali Ibn Abu Talib AB, in March 1991, #07019, #09043, and #09051.

IrAF Serial No.	Type	Delivery	Remarks
6795	Su-25K	1987	Pre-1988 serial, No.114 Sqn
25571	Su-25K	1985	c/n 07021?
25585	Su-25K	1985	Flown to Iran, 1991
25590	Su-25K	1985	Flown to Iran, 1991
25591	Su-25K	1985	Destroyed 1991
25603	Su-25K	1986	Flown to Iran, 1991
25604	Su-25K	1986	Flown to Iran, 1991
25606	Su-25K	1986	Flown to Iran, 1991
25609	Su-25K	1986	Found near Tammuz, 2003
25611	Su-25K	1987	Flown to Iran, 1991
25612	Su-25K	1987	Found near Tammuz, 2003
25613	Su-25K	1987	Flown to Iran, 1991
25615	Su-25K	1987	Flown to Iran, 1991
Unknown	Su-25K	1987	c/n 10306, wreck at Tammuz

Appendix

I. Bibliography

- *History of the Iraqi Armed Forces, Part 17; The Establishment of the Iraqi Air Forces and its Development.* Iraq: Iraqi Ministry of Defence, 1988. (no ISBN; available in Iraq only)

- Mustafa, Genral H. *The June War, 1967, Part II"*, Lebanon: Establishment for Arab Studies and Publication, 1970. (no ISBN available)

- Group of authors *The Role of the Iraqi Armed Forces in the October 1973 War.* Lebanon: Establishment for Arab Studies and Publication, 1974.

- *Memoirs of the Iraqi Air Force CO, Major General K.K. Omar.* Iraq: unpublished document

- *Iraqi Air Force Aircraft flown to Iran, in 1991.* Iraq: Letter from the former Iraqi Ministry of Foreign Affairs to General Secretary UN, issued 2001

- *Various magazines and journals.* Iraq: the Iraqi Air Force and the Iraqi Ministry of Defence, 1970s to 1990s

- *Belgian Military Aircraft.* Belgium: http://belmilac.wetpaint.com

- Personal interviews with various Iraqi Air Force officers, pilots, and ground personnel

- Personal notes from both authors

- Personal notes by Ferenz Vajda, based on original Soviet documents about exports of combat aircraft to Iraq

II. Kit and Decal List

Compiled with kind help of IPMS Austria, the following list includes most of currently available plastic kits, as used for replicating fighters of the Iraqi Air Force.

Table 14: Aircraft Kits to Section

Manufacturer	Subject	Product No.	Remarks
Scale 1:72			
Airfix	Hunter	2073	Old kit requiring plenty of work
Revell	Hunter F.Mk.6	4350	Kit of newest technology with excellent details
Kopro	MiG-17PF	3104 & 3177	30 years old kit with new decals, and still the only with proper proportions
Kopro	MiG-19	3105 & 3178	30 years old kit with new decals and very good proportions
Kopro	MiG-19	3178	
Bilek	MiG-19S	956	Good kit with very good proportions, re-issued with Tally Ho decals (no IrAF)
Bilek	MiG-19P	957	Includes all parts for MiG-19S as well
Bilek	MiG-19PM	958	
Bilek	MiG-21F-13	931	
Revell	MiG-21F-13	4346	Best kit of this version, including recessed panel lines
Bilek	MiG-21PFM	934	
Fujimi	MiG-21PFM		Older kit in need of reworking in many areas, some parts useful for fitment to other kits
Zvezda	MiG-21PFM	7202	Useful kit with few glitches that are easy to correct
Kopro	MiG-21MF	3107	20 years old kit with very good proportions
Kopro	MiG-21bis	7259	
ICM	MiG-21bis	181	Similar to MiG-21s from Kopro, recessed lines, could use some parts from Fujimi
Academy	MiG-23S	1621	Useful for making MiG-23MF
Hasegawa	MiG-23S		Useful for making MiG-23MF
Zvezda	MiG-23MLD	7218	Useful as MiG-23ML
Academy	MiG-27	1654	Useful as basis for MiG-23BN, though some work is required on nose, intakes and engine nozzles
Hasegawa	MiG-27	C10	Useful as basis for MiG-23BN
Zvezda	MiG-27	7228	Useful as basis for MiG-23BN
Condor	MiG-25P		Useful kit, with good fit and correct proportions
Hasegawa	MiG-25P	0434	Older kit, still useful but lacking details
Bilek	MiG-29	920	
Eastern Express	MiG-29	106	
Hasegawa	MiG-29	E11	Older but still useful kit, with very few minor mistakes
Italeri	MiG-29	0184	Older kit
ICM	MiG-29 9-13	141	
Kangnam	MiG-29	7204	

Manufacturer	Subject	Product No.	Remarks
Bilek	Su-17 Fitter G	950	Useful as Su-22M-3K
Bilek	Su-17 Fitter H	936	Including IrAF decals, useful as Su-22M-4K
Bilek	Su-22M-4	970 unknown	
Kopro (Mastercraft)	Su-17/33M-4	KP4125	Useful as Su-22M-4K, recessed panel lines
Kopro (Mastercraft)	Su-20/R	KP4131	Useful as Su-20, recessed panel lines
Kopro (Mastercraft)	Su-22M-3	KP4130	Useful as Su-22M-3K, recessed panel lines
Kopro (Mastercraft)	Su-22M-4/R	KP4128	Including IrAF decals, recessed panel lines, with good selection of weapons
Zvezda	Su-24M	No.7267	
Zvezda	Su-24MR	No.7268	
Revell (ex. Dragon, ex. Zvezda)	Su-24M	No.4399	Including IrAF decals, a very good kit, with good selection of weapons
Zvezda	Su-25	No.7227	Including IrAF decals, recessed panel lines, wide selection of weapons, probably the best Su-25 kit available
Kangnam	Su-25	7201	
Kopro	Su-25	KP3122	Useful kit, including IrAF decals
Heller	Mirage F.1C	80318	Older kit, lacking HF extra dorsal fillet on fin
Hasegawa	Mirage F.1C	B04	Older kit, recessed panel lines, including IFR-probe but lacking HF extra dorsal fillet on fin
Scale 1:48			
Academy	Hunter F.Mk.6	2164	
Academy	Hunter FGA.9	2169	
Hobbycraft	MiG-17F	NC1593	Very simple
Smer	MiG-17F	825	
Smer	MiG-17PF	827	
HiPM	MiG-19S	4809	
Academy	MiG-21PF	2166	
Academy	MiG-21MF	2171	
Italeri (ex-ESCI)	MiG-23M	2649	Only MiG-23 kit in this scale, very good, despite significant problems in
Italeri	MiG-27	2661	
Revell	MiG-25PD	04589	Useful model, with few details that need correction
Academy	MiG-29A	2116	
OEZ	Su-7BMK		
Kopro (Mastercraft)	Su-22M-3	KP3163	
Scale 1:32			
Revell	Hunter F.Mk.6		
Revell	Hunter FGA.9		
Trumpeter	MiG-17F/ PF		
Trumpeter	MiG-19S		
Revell	MiG-21F-13		Very good, incl. metal & resin parts
Revell	MiG-21PF/ PFM	H-267	A bit simple
Revell	MiG-21MF		Lots of weapons
Revell	MiG-29		Needs a lot of work

القوة الجوية الملكية العراقية

Table 15: Decals

Manufacturer	Reference	Content & Commentary
Scale 1:72		
Aero Master	72-127	IrAF Su-22M-4K #22564, out of production
Albatros	ALC-72010	IrAF Mirage F.EQ-5 x2, out of production
Begemot	72015	MiG-25, includes non-existing serial number 35301, still
Carpena	72.02	IrAF F.1EQ-5 #4566, still available
Carpena	72.36	IrAF MiG-29 #29060 (as well as Mi-25 and SA.342K) still available
Cutting Edge	72001	IrAF Su-24 #24624, out of production
ESCI Decals	61	IrAF Hunter T.Mk.66A #567, out of production
Gekko Graphics	72-002	IrAF MiG-21MF #21178, out of production
Fresco	72-034	MiG-29 #29060, out of production
Hi-Decal	72-002	IrAF Su-25K #25590, still available
Hi-Decal	72-005	IrAF MiG-29 #29060, still available
Hi-Decal	72-011	IrAF MiG-21R #21302, still available
Hi-Decal	72-013	IrAF Su-24MK #24346, still available
Hi-Decal	72-015	IrAF MiG-23MS #23047, still available
Hi-Decal	72-017	IrAF MiG-23BN #23181, still available
Hi-Decal	72-028	IrAF MiG-25PDS non-existing #35416, still available
Hi-Decal	72030	IrAF MiG-17F #452, still available
MicroScale/SuperScale	72-607	IrAF full set of national insignia and serial numbers for MiG-21FL and -21bis, MiG-23BN and two Mirage F.1EQs, out of production
Propagteam	72106	IrAF Hunter F.Mk.59 #570, still available
Repliscale	1023	IrAF MiG-29 markings (no serial numbers), still available
Scale 1:48		
Albatros	ALC-48010	IrAF Mirage F.1EQ-5 x2, out of production
Cutting Edge	48001	IrAF Su-24MK #24624, still available
Eagle Strike	48255	IrAF Mirage F.1EQ #4528, still available
Iliad Design	48012	IrAF MiG-15UTI #874, still available
Scale 1:32		
CAM	32A14	Persian Gulf Air Forces Part 1, National Insignia, still available
Eagle Strike	32032	IrAF MiG-21R #21302, still available

Index

A. K. Qasim (General, Iraqi Army) 40
A-Razzaq (Wing Commander, CO IrAF 1963) 25, 27

Dassault Super Etendard 108, 115

Gorky (today Nizhni Novogorod) 46, 51

Hameed Sha'ban (General, CO IrAF 1976-1977, 1984-1990) 17
Hassan al-Khither (1st Lieutenant, later General and Deputy CO IrAF) 30

Ihsan Shudrom (Flight-Lieutenant, later Chief of Staff RJAF) 29
IrAF Air Bases
 – Abu Ubaida al-Jarrah 13, 53, 73, 84, 119, 141
 – al-Assad 13, 62, 83, 84, 87
 – al-Baghdadi 13
 – al-Bakr (Bakr) 13, 72, 76-80, 88, 129, 137, 138
 – al-Hurrya (Hurrya) 13, 53, 123- 125, 128, 130, 132
 – al-Rashid (Rashid) 13, 15, 17, 19, 21, 28, 39, 46-49, 50, 53, 73, 123
 – al-Wahda (Wahda) 13, 53, 54, 61, 73, 96, 101, 108, 118, 125, 131, 141
 – al-Wallid (see also H-3) 13, 28, 29, 42, 50, 70, 73, 137, 138, 144
 – Ali Ibn Abu Talib 13, 56, 60, 66, 73-76, 84, 132, 133, 141
 – Artawi 13
 – Dulu'ya 13
 – Firnas 13, 53, 69
 – H-3 (see also al-Wallid) 13, 73, 138
 – Jaliba 13, 91, 141
 – Qadessiya 13, 67, 69, 70, 79, 86, 88, 96, 125
 – Sa'ad (see also H-2) 13
 – Saddam 13, 96, 99, 107
 – Tammuz 13, 42, 69, 70, 73, 75, 79, 81-85, 87-91, 94, 135, 136, 141-146
IrAF Units
 – Academy 17, 40, 53, 125
 – No.1 Squadron 12, 19, 39
  and Su-7 117-119
  and Su-20 123-125, 133
 – No.5 Squadron
  and Vampire 15-19
  and Venom 21
  and MiG-17 39-40
  and Su-7 118-119
  and Su-20 125, 128-130
 – No.6 Squadron
  and Venom 17
  and Hunter 24-31, 33, 40, 47, 50
  and MiG-29 89-91
 – No.7 Squadron 27, 38, 40-43, 48
 – No.8 Squadron
  and Su-7 118-119
  and Su-20 137-138
 – No.9 Squadron 27, 46-51, 53, 54, 58, 61-62, 65, 67, 69, 101
 – No.10 Squadron 137
 – No.11 Squadron 49-54, 58, 60, 67
 – No.17 Squadron and OCU 50-54, 67
 – No.29 Squadron
  and Hunter 28, 30-33
  and MiG-23 73-75
 – No.39 Squadron 71-75
 – No.44 Squadron 124-130
 – No.47 Squadron 53
 – No.49 Squadron 73-75
 – No.59 Squadron and OCU 72, 75
 – No.63 Squadron 76
 – No.69 Squadron 125, 129
 – No.70 Squadron 53, 56, 81

- No.73 Squadron 75-76
- No.79 Squadron 96, 101, 107, 108
- No.81 Squadron 100, 102, 108-110
- No.84 Squadron 81-87
- No.89 Squadron 107
- No.91 Squadron 101, 107, 108
- No.96 Squadron 81-87
- No.109 Squadron 125-132
- No.114 Squadron 143-146
- No.115 Squadron 141-143
- No.116 Squadron 141-143
- Hunter OCU 25
- Flight College 17, 40, 47
- Flying Leaders School 30, 32-33

Israel
  and 1967 War 12, 13, 24-34, 50, 67
  and 1973 War 42, 50, 51, 63, 119
Izzat (Major, Hunter-pilot, IrAF) 33

Khaled Sarah (Major, MiG-17- & MiG-19-pilot, IrAF) 39, 46

Lugovaya 71

Mohammad Jassem (1st Lieutenant, Hunter-pilot, later General, CO IrAF 1979-1984) 17
Mufeed Saeed (1st Lieutenant, Hunter-pilot) 25
Munthir al-Windawi (1st Lieutenant, Hunter-pilot) 47

Najdat al-Naqeeb (1st Lieutenant, Hunter-pilot, later General and Deputy-CO IrAF) 30
Namiq Saed-Al (Major, MiG-21-pilot) 58
Natiq (Colonel, Hunter-pilot, IrAF) 30-31
Nimaa Dulaymee (1st Lieutenant, Hunter-pilot, later General, CO IrAF 1974-1976) 17

RQ-1 Predator (US UAV) 87

Shehab al-Qaisy (Lieutenant-Colonel, MiG-19 & Su-7-pilot) 41-42
Saddam Fedayeen Insignia 96
Saif-ul-Azam (Flight-Lieutenant, PAF) 29, 32, 33
Shenyang F-7B 55, 60

Weapons & Equipment
  - Almaz 23 (radar) 53
  - AM.39 Exocet 104-114
  - AS.30 & AS.30L 110-114
  - AS.37 (ARM) 96, 101-102
  - Atlis (laser-marker) 102-110
  - Belouga (BLG-66 CBU) 98-101
  - CB-500 (CBU) 142
  - Cyrano IV (radar) 96-111
  - FAB-500T 81-87
  - Kh-23 (AS-7 Kerry) 73, 127
  - Kh-25 (AS-10 Kegler) 129-132
  - Kh-28 (AS-9 Kyle) 128-132
  - Kh-29 (AS-14 Kedge) 126-129
  - Kh-66 50
  - KKR-1 (recce pod) 124
  - Klyon (laser rangefinder) 129
  - MIM-23 HAWK (SAM) 28, 128, 132, 139
  - N003E (Sapfir-23ML radar) 75
  - N019 (Rubin radar) 91
  - OEPS-29 KOLS (IR-system) 91
  - ORO-57K (rocket pod) 39, 45
  - R-2L (radar) 50
  - RP-2U (radar) 46
  - RS-2US (AA-1 Alkali) 46
  - R-3S (AA-2 Atoll, AAM) 49-58, 61-62, 71-72, 75
  - R-13M (AA-2 Atoll, AAM) 53, 58, 60-62, 72-73, 75
  - R-23 (AA-7 Aphex, AAM) 74-75
  - R-24 (AA-7 Aphex, AAM) 74, 76, 78
  - R-27 (AA-10 Alamo, AAM) 91, 93
  - R-40 (AA-6 Acrid, AAM) 81-87
  - R-60 (AA-8 Aphid, AAM) 53, 59, 61, 74-76, 83, 91, 129, 138, 143
  - R.550 Magic (AAM) 98, 101, 103
    on MiG-21 53, 58
  - Remora (ECM pod) 77, 100-109
  - RP-21 (radar) 50
  - SAMP (GP-bombs) 100-110
  - Sirena 2 (RWR) 39, 71, 119
  - SPO-2/-3/-15 (RWR) 53, 77-78
  - SPS-141 124, 128-132
  - SRO-1/-2 (IFF) 49-52, 119
  - Super 530 (Super 530F-1, MRAAM) 96-104
  - Super 530D (MRAAM) 106, 110
  - Syrel (ELINT-pod) 101, 103, 110
  - TP-23 (IRST system) 75
  - TP-26 (IRST system) 75, 83
  - UV-16-57 (rocket pod) 49-50, 73, 119
  - UV-32-57 (rocket pod) 54, 55, 61, 125-126, 129, 142-143

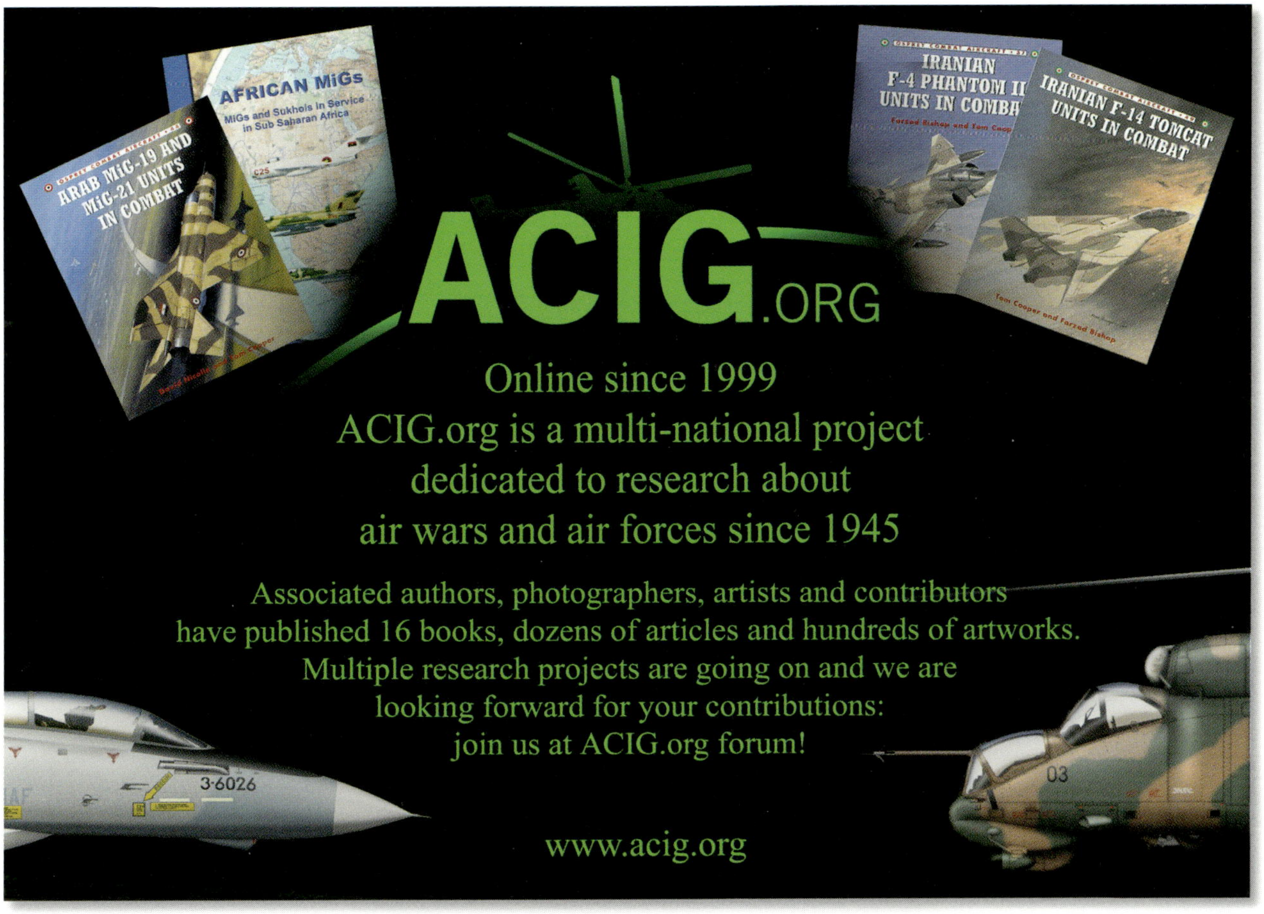
AFRICAN MiGs
MiGs and Sukhois in Service in Sub Saharan Africa
ARAB MiG-19 AND MiG-21 UNITS IN COMBAT
IRANIAN F-4 PHANTOM II UNITS IN COMBAT
IRANIAN F-14 TOMCAT UNITS IN COMBAT
ACIG.ORG
Online since 1999
ACIG.org is a multi-national project
dedicated to research about
air wars and air forces since 1945
Associated authors, photographers, artists and contributors
have published 16 books, dozens of articles and hundreds of artworks.
Multiple research projects are going on and we are
looking forward for your contributions:
join us at ACIG.org forum!
www.acig.org

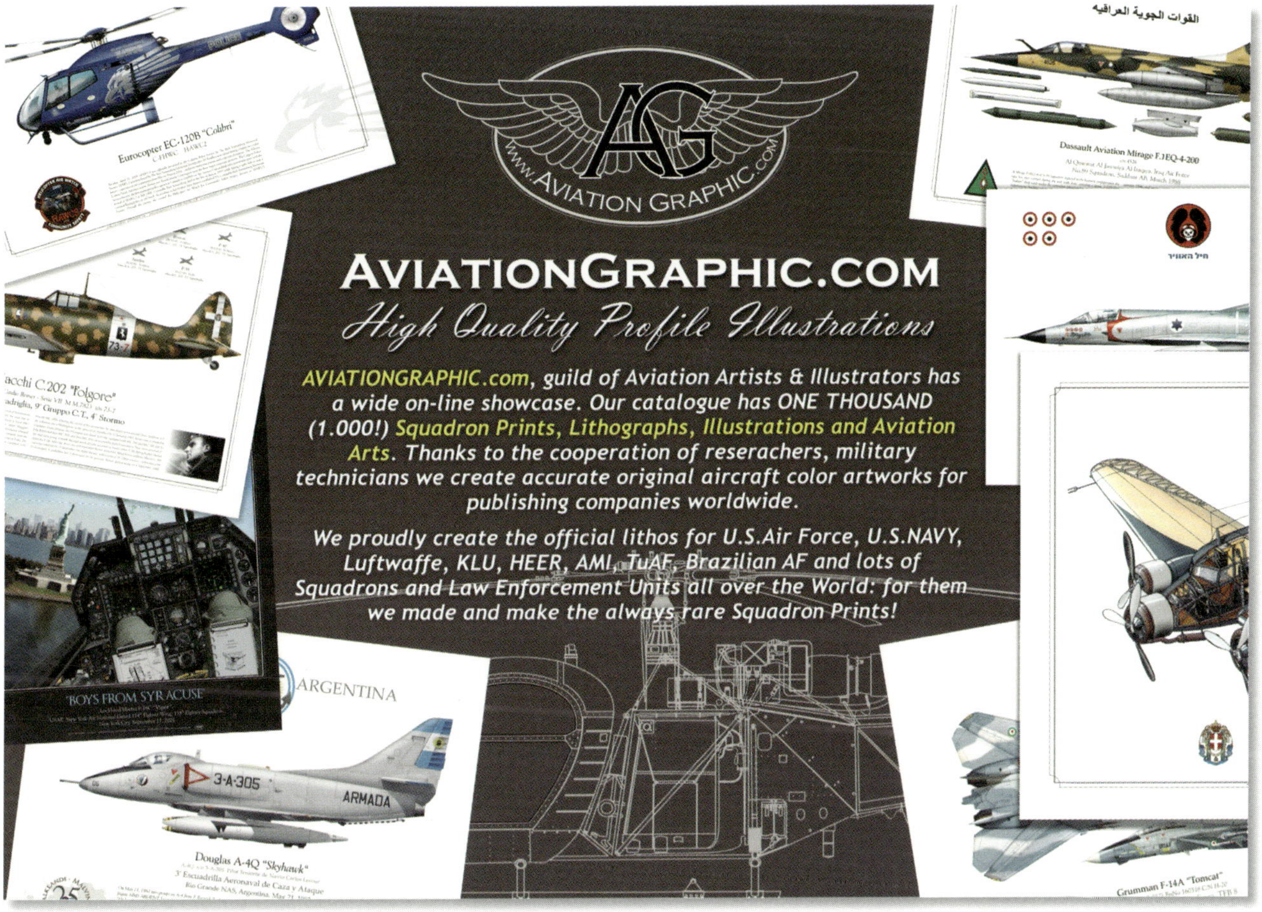
WWW.AVIATION GRAPHIC.COM
AVIATIONGRAPHIC.COM
High Quality Profile Illustrations
AVIATIONGRAPHIC.com, guild of Aviation Artists & Illustrators has a wide on-line showcase. Our catalogue has ONE THOUSAND (1.000!) Squadron Prints, Lithographs, Illustrations and Aviation Arts. Thanks to the cooperation of reserachers, military technicians we create accurate original aircraft color artworks for publishing companies worldwide.
We proudly create the official lithos for U.S.Air Force, U.S.NAVY, Luftwaffe, KLU, HEER, AMI, TuAF, Brazilian AF and lots of Squadrons and Law Enforcement Units all over the World: for them we made and make the always rare Squadron Prints!
Eurocopter EC-120B "Colibri"
Dassault Aviation Mirage F.1EQ-4-200
'BOYS FROM SYRACUSE'
ARGENTINA
3-A-305
ARMADA
Douglas A-4Q "Skyhawk"
Grumman F-14A "Tomcat"

US/British Name	Official IrAF Name	Remarks
Al-Assad	Qadessiya AB (former al-Baghdadi AB)	Former water-place Ain al-Assad, named after a battle in which the Arabs defeated ancient Persians, circa. 600AD
Al-Taqaddum	Tammuz (or Tahmouz) AB	Named after the Babylonian month of Tammuz, 7th month in the year
Balad	al-Bakr AB	Named after former Iraqi President, Ahmad Hassan al-Bakr
Balad SE	Dulu'ya	Dispersal facility, named after geographic location
H-2	Sa'ad AB	Named after an ancient Islamic warrior, from cca. 600AD
H-3	al-Wallid AB	Named after an ancient Islamic warrior, Khalid bin al-Wallid, from circa. 600AD
Jaliba	Jaliba	Dispersal facility
Kirkuk	al-Hurriyah AB	Means "Liberty" in Arabic
Mosul	Firnas AB	Named after a legendary middle-age Andalusian Arab, who attempted to fly using feathered wings
Mudaysis	Talha AB	Dispersal facility named after ancient Islamic warrior
Rashid	al-Rashid AB	Named after Islamic Caliph under which Baghdad prospered
Shoaibah (RAF Shaibah)	al-Wahda AB	Means "Unity" in Arabic
Kut or Kut al-Hayy	Abu Ubaida al-Jarrah AB	Named after an Islamic warrior from circa. 600AD
Tallil	Ali Ibn Abu Talib AB	Named after an Islamic warrior from circa. 600AD
Ar Rumaila SW	Artawi airfield	Named after geographic location
Qayyarah West	Saddam AB	Named after former Iraqi President